MEDIA MANUALS

# Motion Picture Camera Techniques

MEDIA MANUALS

THE ANIMATION STAND
Zoran Perisic

BASIC FILM TECHNIQUE
Ken Daley

CONTINUITY IN FILM & VIDEO
Avril Rowlands

CREATING SPECIAL EFFECTS FOR TV & FILMS
Bernard Wilkie

EFFECTIVE TV PRODUCTION
Gerald Millerson

MOTION PICTURE CAMERA DATA
David W. Samuelson

MOTION PICTURE CAMERA & LIGHTING EQUIPMENT
David W. Samuelson

MOTION PICTURE CAMERA TECHNIQUES
David W. Samuelson

MOTION PICTURE FILM PROCESSING
Dominic Case

16 mm FILM CUTTING
John Burder

SOUND RECORDING AND REPRODUCTION
Glyn Alkin

SOUND TECHNIQUES FOR VIDEO & TV
Glyn Alkin

TV LIGHTING METHODS
Gerald Millerson

THE USE OF MICROPHONES
Alec Nisbett

USING VIDEOTAPE
J. F. Robinson & P. H. Beards

VIDEO CAMERA TECHNIQUES
Gerald Millerson

THE VIDEO STUDIO
Alan Bermingham, Michael Talbot-Smith,
Ken Angold-Stephens & Ed Boyce

YOUR FILM & THE LAB
L. Bernard Happé

# Motion Picture Camera Techniques

## Second Edition

David W. Samuelson

**FOCAL PRESS**
London & Boston

**Focal Press**
is an imprint of Butterworth Scientific

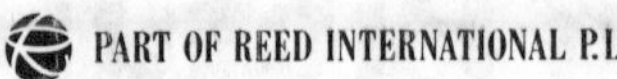




First published 1978
Second edition 1984
Reprinted 1987
Reprinted 1989



**British Library Cataloguing in Publication Data**

Samuelson, David W.
Motion picture camera techniques.—2nd ed.
–(Media manuals)
1. Moving-picture cameras
I. Title II. Series
778.5'3 TR880

ISBN 0-240-51247-2

**Library of Congress Cataloging in Publication Data**

Samuelson, David W.
Motion picture camera techniques

Bibliography: p.
1. Cinematography. I. Title.
TR850.S315 1984 778.5'3 84-6151
ISBN 0-240-51247-2

Composition by Genesis Typesetting, Borough Green, Sevenoaks, Kent
Printed and bound by Hartnolls Ltd, Bodmin, Cornwall

# Contents

# Introduction

Without a doubt, the most important ingredient of a successful film is a good screenplay. Only after this does the application of men, money and machinery to translate the pages of a script into scenes in a film become significant.

Film making combines many aspects. It is creative and yet is bound by technology, standards and tradition. Individuals make contributions which leave an indelible stamp upon the production, and yet it is essentially a team effort. Masterpieces were made with the equipment of yesteryear, yet new equipment, new materials and new techniques make new concepts, previously beyond the imagination, possible to achieve. There are new limits—but even these rules are there to be broken. People can make good names for themselves, reputations may be ruined by the maladroitness of others and, ironically, the most important person to be satisfied may have nothing to do with the project in hand but be the man who is prepared to finance a future production on the strength of the current one.

Cinematography is an art; of that there is no doubt. But the canvas is expensive and unless commercial viability is observed at all stages, he who is prodigal or avaricious faces the possibility of a lifetime of frustration. There is no-one more sorrowful than an unemployed film maker, especially one who has exceptional talent and motivation but whom no-one can afford to employ. It must never be forgotten that films are usually made with some one else's money and respect for this may well affect an individual's continuity of employment.

Rarely is there an opportunity for undisciplined experiment for, invariably, there is too much money at stake, putting expertise at a premium. A bold imagination, though the very life blood of cinema and necessary for its survival, may only be indulged if the results are virtually guaranteed, for failure is too costly to contemplate.

What then should be the aim? What is good and what is worthwhile?

If the end result satisfies only its creators it is merely an ego trip. If it succeeds in recouping its cost and feeds, clothes and warms those who work on it and even makes a profit for re-investment in another film and regeneration of the film industry then it is successful indeed. Film makers should never underestimate the importance of money or lose their respect for it, particularly if it belongs to another person.

# Acknowledgments

When someone writes such a book as this, it is important for him to have friends from whom he may seek advice or verification of facts, or simply chat to, to help sort out his own thoughts on a particular aspect of film making.

Among those upon whom I have imposed while writing this, and my previous book, I would particularly like to thank the following:

My brothers and colleagues at Samuelson Group PLC;

Freddie Young, OBE, BSC, three times Oscar-winning Director of Photography;

Jack Hildyard, BSC, Ossie Morris, BSC (and his crew), and the late Geoff Unsworth, BSC, all Oscar-winning Directors of Photography.

Robert E. Gottschalk, President of Panavision;

Herb Lightman, Editor of American Cinematographer Magazine;

D. Kimbley, of Kodak Motion Picture Division;

Tony Lumkin, Technical Director of EMI Studios;

Bernard Happé, a mine of all technical information;

Bill Pollard who designs Depth of Field Calculators;

Geoff Smith of Southern Lighting Associates who understands lighting;

Steven Valentine and Barry Orchard of Samfreight whose business is shipping and carnets;

Roland Chase of Colour Film Services, who knows 16mm film;

John Corbett and C. B. B. Wood of the BBC;

Dr P. N. Cardew of St Mary's Hospital;

Les Ostinelli of Technicolour Limited, truly a cameraman's father confessor;

Vic Margutti formerly of Rank (Denham) Laboratories Limited, master of travelling mattes;

Ed di Gulio and Milt Forman of Cinema Products Limited;

Bill Couch who advises about insurance;

Herbert Raditschnig who taught me how to use ropes;

Peter Samuelson, now resident in the USA, who advised on American terminology;

Rex Ebbetts of Filmatic for advice about the usage of filmstock;

Brian Salt for helping me compile the sun tables;

The BBC for permission to publish the HMI tables;

Roy Field and R. A. Dimbleby who explained the complexities of colour differential travelling mattes;

Barney Fishbien, Peter Govey, Harry Hart, Pat Hayes, Dennis Holland, Chris Powney, Dennis Robertson, Bob Stern, Jim Webb and Leslie Wheeler all of whom gave advice and material for artwork;

The EBU for information about the time base code;

Ousama Rawi, Harry Waxman and colleagues in the British Society of Cinematographera;

My colleagues in the British Kinematograph, Sound and Television Society and the Association of Cinematograph, Television and Allied Technicians;

Fellow delegates at UNIATEC, the International Society of Cinema Technical Societies;

And three people in particular who made me want to write this book and made it possible, my late father, G. B. Samuelson, who always told me to be a technician, never a producer, my mother who is a matriarch, if ever there was one and Elaine.

While most of the artwork in this book has been derived from original photographs taken by the author, or supplied by Samuelson Film Service Limited, some are based on catalogues and descriptive literature and the author wishes to thank the following companies for their co-operation in this respect.

A.T.A; Arnold & Richter; Autocue; Bell & Howell; British Movietone News; Century; Cinema Products; FPA; GSE; Hollywood Film Company; Image Devices; Kodak; Kudelski; Motion Picture Research; Neilson Hordell; Northfield Appliances; Oxberry; Paillard Bolex; Panavision; Redlake; Seiki; Sylvania; Technicolor.

Over the years I have been an avid reader and collector of film technical books and if I were to compile a bibliography of all the reference books which have contributed to my knowledge of film technology it would be very long and, inevitably, many would be omitted out of sheer forgetfulness. It is better to mention none but to thank all those authors who over the years have kept me informed or encouraged me to gain further experience, which is the best teacher of all. Having said that, I would like to make the exception and give special credit and thanks to the Editors of the American Cinematographer Magazine and the American Cinematographer Manual (and its predecessor the Jackson Rose Manual) which have been my constant sources of reference since the first day I ever used a motion picture camera professionally.

# The Script

At an early stage in planning a production, a script should have been written and agreed upon. It is the key document in making a film and without a good script or screenplay there can be little hope of a successful and financially viable production.

All senior personnel on a production are given a copy of the script. In many cases, they have read this before agreeing to become involved. Once the production is under way it is often advantageous for other people to have copies. This gives them an added feeling of importance and incentive. If they have an opportunity to study it prior to each day of shooting, they come to work with a knowledge of what they have to do and why. Additionally, it gives all concerned the maximum time to think about any specialised equipment, effects and props that may be required and to ensure that they are to hand when called for.

The 'final' script is not a sacrosanct document that cannot be altered. If any suggestions and contributions are made which are worth including, this practice should be encouraged, no matter from whom the suggestions come. Film-making is a team effort.

A better idea of the timing of scenes emerges as a production proceeds. So it may be possible and advantageous to tighten up a script, to expand ideas that work well, and possibly delete certain parts. There is no point in spending money on a 'first cut' which lasts four hours if it must subsequently be edited down to one hour 40 min. One of the advantages in shooting a script in sequence is that modifications may be more easily incorporated and are less likely to involve re-shooting.

## The script and the director of photography

When the director of photography reads his script prior to starting a production he should be able to visualise individual shots from it. After discussion with the director, he should have an idea of the overall 'look' that the film should have, and to plan accordingly. He breaks down scenes and gives particular thought to any which require special effects, opticals or process shots. He plans how to cope with any known artist's fetishes which may occur (eg. from which side they like to be seen and key lit) and generally plans the continuity of mood and effect.

If he can be mentally four script pages ahead of the shot in hand he may well be able to save the production much time and money.

47 EXT. NIGHT

Light shines through the windows of the master bedroom, the rest of the house in darkness. From his hiding place in the grounds TROUT anticipates a glimpse of LUCETTE preparing for bed.

48 INT. NIGHT LUCETTE'S BEDROOM

The radio plays quietly from a corner. LUCETTE, still dressed, seated behind a small antique desk, more decorative than practical. Laid out before her are D'Arcy's personal effects; driving gloves, shirt, jacket, watch, wallet, Leica, and six or seven photographs which she examines. Flipping through them quickly; a couple of 'Art' shots (flowers, trees) -- the rest of BEN; one particularly attracts her interest.

[illegible]graph of BEN[illegible] her [illegible] CAMILLA

## IMPLICATIONS IN THE SCRIPT

A few descriptive words in the script may mean a good deal of preplanning for the director of photography.

# Going on a Recce

If a film project involves an important element of shooting on location the director of photography will undoubtedly be asked to join the director and others in reconnoitring the proposed venues long before the main unit, artists and equipment are called. Time and money spent on careful and thorough local pre-planning is effort well spent, even though it may involve several days of work locally or even travelling half way round the world.

## What the cameraman must look for

Before going on a recce the cameraman must read the script and discuss its interpretation with the director. In visiting the location he can plan the means rather more exactly and endeavour to foresee what will be required in its realisation. The selected locations for a particular film must not only suit the script but be logistically possible and the most economical choice, all other things considered.

Having agreed to the venue the cameraman may begin to visualise the camera set-ups and lighting requirements. He will decide if his normal equipment complement is sufficient, if some can be left behind to reduce transportation costs or if augmentation is required—more sophisticated dollys and cranes, a smaller lighter camera, wider aperture or longer or shorter focal length lenses, a cherry picker or scaffold towers to elevate cameras or lights, and so on.

The lighting requirements need particular attention. The cameraman must assess what is the largest area that must be artificially illuminated at one time, think out if it can be reduced, calculate how many of each type of light is required, and how much cabling. If an electrician is not also sent on the recce, the cameraman must judge if the local power supply is suitable or if a generator or alternator will be required. A generator must have a parking place not too far from the point of shooting, (to minimise voltage drop) but not close to domestic bedrooms if the shoot is at night.

Other items to be considered on a recce include obtaining local permits, fitting the schedule to desired weather conditions, and watching out for TV aerials and other items, incongruous in period settings.

A compass, a still camera or even a Super 8 camera or portable VTR are also sometimes taken.

LOCATION ASSESSMENT

When making a location assessment reconnaissance be very thorough in judging the possibilities of every likely site and keep detailed notes for subsequent recall. 1. Using a director's viewfinder to visualise a camera set-up. Notes may be taken on a pad (2) or on a portable tape recorder (3). Still photographs may be taken with an instant camera (4) or by a more conventional means (7). Moving pictures may be taken on Super 8 (6) or on a portable VTR (5).

# Planning the Day

When filming scripted action by sunlight it is often advisable to plan the day's shooting schedule to take advantage of the direction, height and intensity of the sun.

## Direction of sun

Directors of photography very often prefer to use powerful artificial lighting (brutes, minibrutes, HMI lights or reflectors) to model the faces of the principal artists and use available sunshine as backlight (contre-jour). By this arrangement the most important aspect of lighting, modelling, is completely under control. In addition, the strong backlight separates artists and objects from the background and gives character to the background lighting. Where it is decided to shoot against the sun as much as possible, the director, the cameraman and others concerned should consider the script in advance of shooting and plan a day's camera set-ups accordingly. When looking at locations it is sometimes useful to have a compass to note the direction that immovable features face. Another reason for considering the direction of the sun in advance is to bring out the character of a location. The relationships between buildings, and objects and their textures, especially of rough stone and rocks, is greatly affected by the direction of the sun. Landscapes too, especially where there are hills and trees, may well look much more interesting at one end of a shooting day than the other.

## Height of sun

A high sun can be the worst possible lighting situation for filming. Artists' eyes became dark sockets without detail, their nose and chin shadows become long and obtrusive. People tend to become merged with backgrounds and the backgrounds themselves are 'flattened out'. Some locations, such as sand dunes and other low undulating landscapes can only be interestingly photographed in the first and last hours of the day.

## Intensity of sun

The so called 'magical minutes' just after daybreak and before dusk are often used for filming 'day for night' sequences. During this period the sun has little strength compared with practical lamps, such as street and vehicle lights and illuminated windows, which, as a consequence stand out boldly.

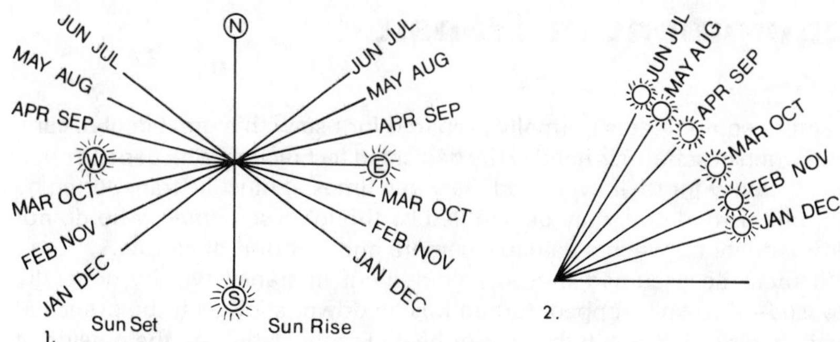


## SUN POSITIONS

1. Extreme positions of the sun (50°N), for southern hemisphere transpose by 6 months; 2. Maximum height of sun (50°N).

### SUNRISE AND SUNSET POSITIONS

North or South of due East (sunrise) or due West (sunset)

| Lat. | mid. Jan. | mid. Feb. | mid. Mar. | mid. Apr. | mid. May | mid. June | mid. July | mid. Aug. | mid. Sept. | mid. Oct. | mid. Nov. | mid. Dec. |
|---|---|---|---|---|---|---|---|---|---|---|---|---|
| 60°N | 9°S | 17°S | 26°S | 40°S | 49°S | 53°S | 51°S | 44°S | 33°S | 21°S | 12°S | 7°S |
| 55°N | 14°S | 22°S | 31°S | 45°S | 54°S | 58°S | 56°S | 49°S | 38°S | 26°S | 17°S | 12°S |
| 55°N | 19°S | 27°S | 36°S | 50°S | 59°S | 63°S | 61°S | 54°S | 43°S | 31°S | 22°S | 17°S |
| 40°N | 29°S | 37°S | 46°S | 60°S | 69°S | 73°S | 71°S | 64°S | 53°S | 41°S | 32°S | 27°S |
| 30°N | 39°S | 47°S | 56°S | 70°S | 79°S | 83°S | 81°S | 74°S | 63°S | 51°S | 42°S | 37°S |
| 20°N | 49°S | 57°S | 66°S | 80°S | 89°S | 87°N | 89°N | 84°S | 73°S | 61°S | 52°S | 47°S |
| 10°N | 59°S | 67°S | 76°S | 90°S | 81°N | 77°N | 79°N | 86°N | 83°S | 71°S | 62°S | 57°S |
| 0° | 69°S | 77°S | 86°S | 80°N | 71°N | 67°N | 69°N | 76°N | 87°N | 81°S | 72°S | 67°S |
| 10°S | 79°S | 87°N | 84°N | 70°N | 61°N | 57°N | 59°N | 66°N | 77°N | 89°N | 82°S | 77°S |
| 20°S | 89°S | 83°N | 74°N | 60°N | 51°N | 47°N | 49°N | 56°N | 67°N | 79°N | 88°N | 87°S |
| 30°S | 81°N | 73°N | 64°N | 50°N | 41°N | 37°N | 39°N | 46°N | 57°N | 69°N | 78°N | 83°N |
| 40°S | 71°N | 63°N | 54°N | 40°N | 31°N | 27°N | 29°N | 36°N | 47°N | 69°N | 78°N | 83°N |
| 50°S | 61°N | 53°N | 44°N | 30°N | 21°N | 17°N | 19°N | 26°N | 37°N | 49°N | 58°N | 63°N |
| 55°S | 56°N | 48°N | 39°N | 25°N | 16°N | 12°N | 14°N | 21°N | 32°N | 44°N | 53°N | 58°N |
| 60°S | 51°N | 43°N | 34°N | 20°N | 11°N | 7°N | 9°N | 16°N | 27°N | 39°N | 48°N | 53°N |

### HOURS OF DAYLIGHT

| Lat. | mid. Jan. | mid. Feb. | mid. Mar. | mid. Apr. | mid. May | mid. June | mid. July | mid. Aug. | mid. Sept. | mid. Oct. | mid. Nov. | mid. Dec. |
|---|---|---|---|---|---|---|---|---|---|---|---|---|
| 60° | 6¾ | 9¼ | 11¾ | 14½ | 17¼ | 18¾ | 18 | 15¾ | 13 | 10¼ | 7½ | 6 |
| 50° | 8½ | 10¼ | 12 | 13¾ | 15½ | 16¼ | 16 | 15½ | 12½ | 10¾ | 9 | 8 |
| 40° | 9¾ | 10¾ | 12 | 13¾ | 14½ | 15 | 14¾ | 13¾ | 12½ | 11¼ | 10 | 9¼ |
| 30° | 10½ | 11¼ | 12 | 13 | 13¾ | 14 | 14 | 13¼ | 12¼ | 11½ | 10½ | 10¼ |
| 20° | 11 | 11½ | 12 | 12½ | 13 | 13¼ | 13¼ | 12¾ | 12¼ | 11¾ | 11¼ | 11 |
| 10° | 11½ | 11¾ | 12 | 12¼ | 12½ | 12¾ | 12¾ | 12½ | 12¼ | 12 | 11¾ | 11½ |
| 0° | 12 | 12 | 12 | 12 | 12 | 12 | 12 | 12 | 12 | 12 | 12 | 12 |
| 10° | 12¾ | 12½ | 12¼ | 12 | 11¾ | 11½ | 11½ | 11¾ | 12 | 12¼ | 12½ | 12¾ |
| 20° | 13¼ | 12¾ | 12¼ | 11¾ | 11¼ | 11 | 11 | 11½ | 12 | 12½ | 13 | 13¼ |
| 30° | 14 | 13¼ | 12¼ | 11½ | 10½ | 10¼ | 10½ | 11¼ | 12 | 13 | 13¾ | 14 |
| 40° | 14¾ | 13¾ | 12½ | 11¼ | 10 | 9¼ | 9¾ | 10¾ | 12 | 13¼ | 14½ | 15 |
| 50° | 16 | 15½ | 12½ | 10¾ | 9 | 8 | 8½ | 10¼ | 12 | 13¾ | 15½ | 16¼ |
| 60° | 18 | 15¾ | 13 | 10¼ | 7½ | 6 | 6¾ | 9¼ | 11¾ | 14½ | 17¼ | 18¾ |

*Travel is an essential component of film making.*

# Equipment in Transit

Camera equipment is normally prepared for use with a great deal of care, and when in action it is handled by dedicated technicians who depend upon its efficiency for their livelihood. Between times, it must be transported by air, rail or road and, may be handled by thoughtless people, who do not differentiate between a valuable camera and a carton of carrots.

It must be assumed that any equipment in transit will be generally abused—thrown, dropped, turned upside down, allowed to be drenched with water or left out in the cold or heat. 'Fragile' labels on the outside of cases are of little avail.

## Packaging for transit

Equipment must be cased with abuse in mind. Ideally, camera cases should combine strength with lightness, convenience of handling and even good looks. Traditional wooden cases are too heavy for economical transport by air and on location, tend to be left behind through sheer fatigue or laziness. Modern lightweight cases, by comparison, are more likely to be kept on hand and always available to protect equipment from the hazards of accident and dirt.

## Labelling

When a unit needs many cases or where it is being shipped abroad and liable to examination by customs officials, a numbered label should be tied on every case and lists issued detailing the contents and number of every case.

## Case identification

When several units are travelling together, identify the cases belonging to different units by say, strips of different coloured camera tapes.

## Shipping lists

When travelling abroad, whether or not a carnet is used (see page 22) a shipping list must be prepared. This list must detail every item in every package together with serial numbers, countries of origin, weights and dimensions. It *must* be accurate. To have an incorrect serial number pointed out by an unfriendly foreign customs officer may be extremely embarrassing.

## Air travel

Undeveloped film likely to be sent by public air transport should be wrapped in special protective material in case it is X-rayed.

EQUIPMENT IN TRANSIT

1. Cases come in many shapes and sizes.

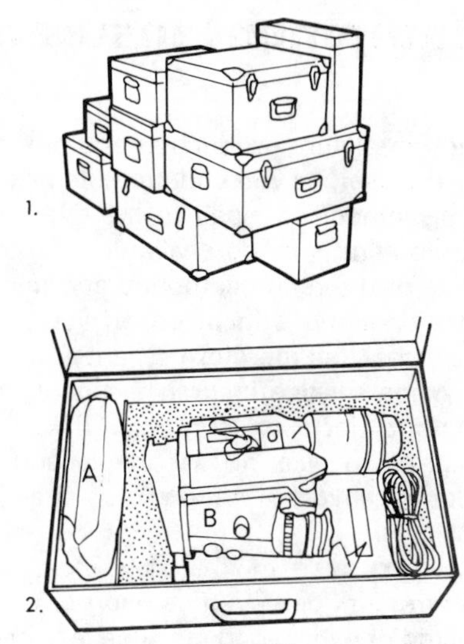


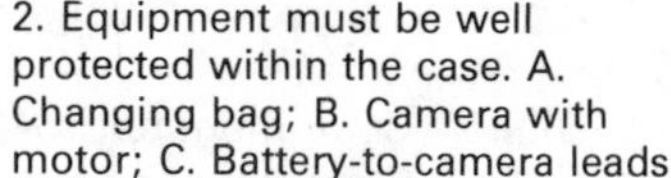

2. Equipment must be well protected within the case. A. Changing bag; B. Camera with motor; C. Battery-to-camera leads.

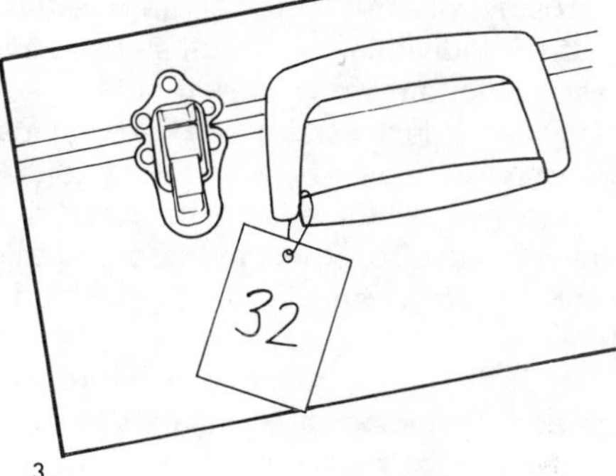


3. Cases on a shipping list should be identified by numbered labels.

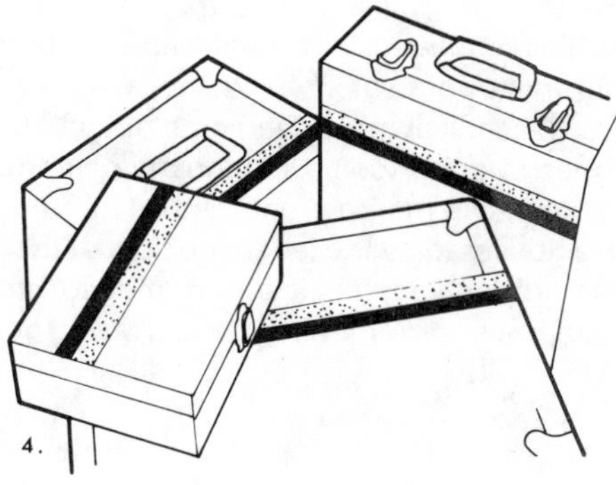


4. Cases belonging to each unit on a multi-unit shoot should be identified by strips of different coloured camera tape.

*No one is rich enough to afford not to insure.*

# Equipment Insurance

Motion picture production equipment is usually expensive and if owned by the user, he will undoubtedly take out an insurance to cover the cost of replacement or repair in the event of accidental loss or damage. Hired equipment is made available in accordance with the conditions of business of the rental company and invariably makes the user liable for any loss or damage incurred, whatever the cause and whatever the consequences. It is therefore essential to insure.

When seeking insurance it is necessary to disclose to underwriters all material facts affecting the risk. Any insurance contract requires the insured to take all reasonable and ordinary precautions to protect the interest covered. One should act as if uninsured.

## Taking care is the best insurance

No prudent person, if uninsured, would leave valuable equipment unprotected or exposed to possible loss or damage. Insurance is really intended to provide for the accidental, unfortuitous and unforeseen event and it is up to the individual to take all steps to avoid loss or damage. If he fails to do so he not only runs a risk that the insurers may not meet his claim, but if they do settle the claim, can find that insurance is difficult, or sometimes impossible to obtain on future occasions. If the equipment is being used in a manner where there is an increased risk of loss or damage, insurers are legally entitled to expect to be told of this beforehand. When a loss does occur the police and the insurance company must be notified immediately.

Whilst the film producer can insure the picture negative against loss or damage, including processing damage or, in certain cases, damage caused by a fault in the camera, an insurer expects that the technician will have carried out tests of the filmstock and the camera and other equipment prior to use. It is customary when starting a feature film to secure sufficient filmstock to use a single batch throughout, or to test each batch number prior to use. Obviously, insurers will not pay for any losses arising from technical misjudgements or similar human errors of the kind sometimes referred to, in the industry's idiom, as 'finger trouble'.

Insurance may be regarded as an 'umbrella', but all insurance policies are governed by the legal principle of the 'utmost good faith'. If, despite all the proper care, it is lost or damaged, steps must be taken to recover the equipment or minimise the loss as if there were no insurance in force.

EQUIPMENT INSURANCE

1, 2. Never leave equipment on show and unattended; 3, 4. If using the equipment under hazardous conditions tell the insurance company beforehand; 5. Test everything before shooting; 6. Insure against all hazards, expected and unexpected.

# Shipping by Carnet

The 'ATA Carnet de Passages en Douane for Temporary Admission' is a convenient customs document, which permits equipment to cross many international borders, to remain there and be used for up to 12 months, provided that every item is exported before the period of validity expires. The ATA carnet is a form of guarantee or bond, made by a local chamber of commerce to the governments concerned, which pledges that if the equipment is not removed, all duties and taxes will be paid in full. Carnets may be used whether the equipment is shipped through brokers or taken by an individual as accompanied baggage.

Before a carnet is issued, a 'shipping list' must be compiled detailing every item of equipment to be transported, together with serial numbers (where applicable), countries of origin, and values. Every package, case or box must be sequentially numbered and the list drawn up detailing sizes and weights. Once drawn up, neither the list nor the goods themselves may be altered. Thus, all the equipment must travel together. A single item may be returned prior to the remainder provided the carnet travels with it, the item is deleted and the carnet returned to travel with the remainder of the equipment. Filmstock, to be returned for processing, should not be on the carnet.

At least two copies of the list must be provided for each border to be crossed. The chamber of commerce will then supply the document with sufficient pages to deposit one copy each time a country is left and another when the next is entered. One person who is travelling with the equipment will be required to sign the carnet and the relevant documents at every customs post. Blue carnet vouchers are used when the equipment is in transit only, white when it is to be used. Failure to observe the rules correctly and to return the carnet to the local chamber of commerce within 19 months of its issue will result in the payment of duty, tax or a penalty.

## Carnet in use

Before travelling, the person responsible for the carnet should verify that there are no inaccuracies or deficiencies in shipping lists, that he knows where to find serial numbers of the various items of equipment and that every digit of every number is correct. All the sheets of the carnet are numbered from one onwards and the *correct pair* must be used for each country visited. When leaving the country where the carnet was raised the local customs officer stamps the first page to confirm that the goods have, in fact, been exported. Upon arrival at the first port of call the customs officer there stamps the next page as evidence that the equipment has been imported, and so on until it is re-imported into the original country and the carnet document is returned to the local chamber of commerce.

The person responsible for the carnet must allow plenty of time at each border, insist that a customs inspection be carried out and the form stamped and never be put off by a wave of the hand throughout the journey.

THE LONDON CHAMBER OF COMMERCE & INDUSTRY
(Issuing Association) ............
*(Association émettrice)*

ATA/GB/LO

INTERNATIONAL GUARANTEE CHAIN
*CHAINE DE GARANTIE INTERNATIONALE*

A.T.A. CARNET No. 96805
*CARNET A.T.A. No.*

CARNET DE PASSAGES EN DOUANE FOR TEMPORARY ADMISSION
CARNET DE PASSAGES EN DOUANE POUR L'ADMISSION TEMPORAIRE

CUSTOMS CONVENTION ON THE A.T.A. CARNET FOR THE TEMPORARY ADMISSION OF GOODS
*CONVENTION DOUANIERE SUR LE CARNET A.T.A. POUR L'ADMISSION TEMPORAIRE DE MARCHANDISES*

(Before completing the carnet, please read notes on page 3 of the cover)
*(Avant de remplir le carnet, lire la notice page 3 de la couverture)*

CARNET VALID UNTIL 28 FEB 1979 INCLUSIVE
*CARNET VALABLE JUSQU'AU* *INCLUS*

ISSUED BY The London Chamber of Commerce & Industry, 69-75, Cannon Street, London, E.C.4N. 5AB.
*DELIVRE PAR*

HOLDER SAMFREIGHT LTD, 303 CRICKLEWOOD BDWY LONDON N.W.. 2
*TITULAIRE*

REPRESENTED BY*) NOEL VERY, ANDREW HOLLAND, MIKE SPELLMAN OR ANY AUTHORISED PERSONS
*REPRESENTE PAR*)*

Intended use of good: /*Utilisation prévue des marchandises* PROFESSIONAL EQUIPMENT

This carnet may be used in the following countries under the guarantee of the following associations: /*Ce carnet est valable dans les pays ci-après, sous la garantie des associations suivantes :*

**AUSTRALIA**
The Melbourne Chamber of Commerce, Melbourne.
**AUSTRIA**
Bundeskammer der Gewerblichen Wirtschaft, Vienna.
**BELGIUM & LUXEMBURG**
Federation Nationale des Chambres de Commerce et d'Industrie de Belgique, Brussels.
**BULGARIA**
Bulgaria Chamber of Commerce, Sofia.
**CANADA**
The Canadian Chamber of Commerce, Montreal 128, Quebec.
**CYPRUS**
Cyprus Chamber of Commerce, Nicosia.
**CZECHOSLOVAKIA**
Ceskoslovenska Obchodni Komora, Prague.
**DENMARK**
Kobenhavns Handelskammer, Copenhagen.
**FINLAND**
The Central Chamber of Commerce of Finland, Helsinki.
**FRANCE**
Chambre de Commerce et d'Industrie de Paris, Paris.
**GERMANY**
Deutscher Industrie-und Handelstag, Bonn.
**GIBRALTAR**
Gibraltar Chamber of Commerce, Gibraltar.
**GREECE**
The Athens Chamber of Commerce and Industry, Athens.
**HONG KONG**
The Hong Kong General Chamber of Commerce
**HUNGARY**
Magyar Kereskadelmi Kamara, Budapest
**ICELAND**
Iceland Chamber of Commerce (Verzlunarrad Islands) Reykjavik
**IRAN**
Iran Chamber of Commerce, Industries & Mines, Teheran.
**IRELAND**
The Dublin Chamber of Commerce, Dublin.
**ISRAEL**
Tel-Aviv-Yaffo Chamber of Commerce, Tel-Aviv.
**ITALY**
Unione Italiana delle Camere di Commercio Industria e Agricoltura Rome
**IVORY COAST**
The Ivory Coast Chamber of Commerce, Inc. Abidjan.
**JAPAN**
Japan Chamber of Commerce, Tokyo.
**NETHERLANDS**
Kamer van Koophandel en Fabrieken voor 'S-Gravenhage, 'S-Gravenhage.
**NORWAY**
Oslo Chamber of Commerce, Oslo.
**POLAND**
Polish Chamber of Foreign Trade, Warsaw.
**PORTUGAL**
Associacao Comercial de Lisboa, Lisbon.
**ROMANIA**
Camera di Comert & Republich Socialiste Romania, Bucarest.
**SOUTH AFRICA**
The Association of Chambers of Commerce of South Africa, Johannesburg.
**SPAIN**
Consejo Superior de las Cameras Oficiales de Comercio Industria Navegacion de Espana, Madrid.
**SWEDEN**
The Stockholm Chamber of Commerce, Stockholm.
**SWITZERLAND**
Alliance des Chambres de Commerce Suisses, Geneva.
**TURKEY**
Union of Chambers of Commerce of Produce Exchanges in Turkey, Ankara.
**UNITED KINGDOM**
The London Chamber of Commerce and Industry, London.
**UNITED STATES OF AMERICA**
U.S. Council of the International Chamber of Commerce Inc., New York.
**YUGOSLAVIA**
The Yugoslav Federal Economic Chamber, Belgrade.

The holder of this carnet and his representative will be held responsible for compliance with the laws and regulations of the country of departure and the countries of importation./*A charge pour le titulaire et son représentant de se conformer aux lois et règlements du pays de départ et des pays d'importation.*

Issued at/*Emis à* London date 1 MAR 1978

(Holders signature/Signature du titulair)

(Signature of authorised Official of the Issuing Association
*Signature du Délégué de l'Association emettrice)*

THE LONDON CHAMBER OF COMMERCE

**TO BE RETURNED TO THE LONDON CHAMBER OF COMMERCE AFTER USE.**

CERTIFICATE BY CUSTOMS AUTHORITIES/*ATTESTATION DES AUTORITES DOUANIERES*

1. Identification marks have been affixed as indicated in column 7 against the following item No(s). of the General List
*Apposés les marques d'identification mentionnées dans la colonne 7 en regard du(des) numéro(s) d'ordre suivant(s) de la liste générale* ............

2. Goods examined.*)/*Vérifié les marchandises.*)*

3. Registered under reference No.*) *Enregistré sous le no.*)*

| (Customs office) *(Bureau de douane)* | (Place/*Lieu*) | (Date/*Date*) | (Signature and stamp) *(Signature et Timbre)* |
|---|---|---|---|

*) Delete if inapplicable./*Biffer s'il y a lieu.*

Typical carnet front cover listing the countries in which a carnet may be used, and the titles of the local chambers of commerce.

*Without light, life and photography would be very dull.*

# The Nature of Light

Light is vital to photography and a cameraman should have some knowledge of its properties.

## Electromagnetic radiation

A pebble thrown into a pool of still water causes a wave or ripple to emanate from the point of contact. Leaves floating on the surface bob up and down and are affected by the wave but not moved away. From this it may be deduced that the water is not moved away from the source of radiation, but vibrates. As a circle of vibration moves away from its point of origin and its energy is spread in an ever-increasing circle, its strength (amplitude) is diminished, but its speed (frequency) remains constant.

If, instead of a single pebble, a plunger is used, moving up and down at regular intervals, it will cause a continuing series of waves all at the same distance apart across the pool (wavelength). If instead of water we have thick oil, the ripples would move more slowly while still remaining the same distance apart. If such oil were blobs in the water in the shape of a lens or prism then, where the oil was thickest, the ripples would be most delayed. Where the ripples meet the oil at an angle the progressive slowing down would change their direction, effectively focusing, dispersing or bending the waves to an extent depending upon the thickness of the oil and the angles of incidence.

A smooth wall would reflect the ripples at an angle equivalent to that at which it was struck by them. A group of rough rocks on the other hand would break up the regular movement and disperse the ripples in all directions.

Light waves (together with radio, X-rays etc.) are electromagnetic waves which conform to the same principles as the pebble in the pond. The wavelength of visible light being around 1/50,000 in. (0.0005 mm).

## Colour

Colour is a sensation in the eye and brain mechanism actuated by different wavelengths of light. These wavelengths are measured in nanometers (nm) or millimicrons (m$\mu$) which are millionths of a millimetre (or the now obsolete Anstrom units—Å; 10 nm or m$\mu$ = 1 Å).

White light is made up of a mixture of all colours in the spectrum, each having its own wavelength—ranging from 400 (violet) to 700 (red) nm or m$\mu$, viz:

| 400 | 475 | 484 | 492 | 508 | 565 |
|---|---|---|---|---|---|
| Violet | Blue-purple | Blue | Blue-green | Green | Green-yellow |

| 578 | 590 | 615 | 700 | nm |
|---|---|---|---|---|
| Yellow | Orange | Orange-red | Red | |

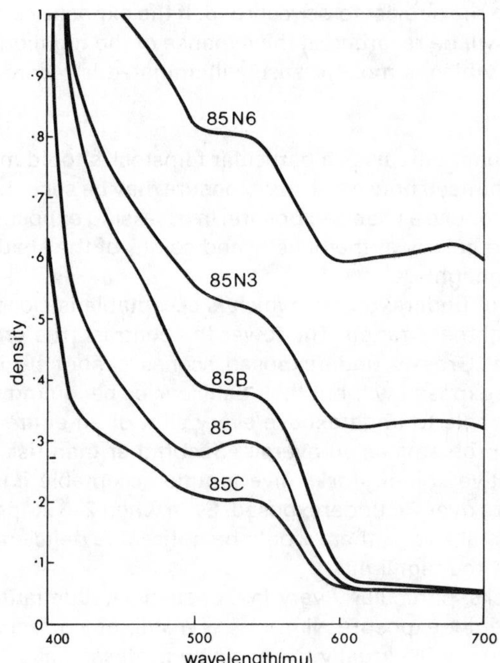


THE NATURE OF LIGHT

The manufacturers of filmstock and colour filters publish graphs to illustrate the sensitivity of their film to certain colours and the effect of their filters. To interpret such graphs (which usually only describe the colours of the spectrum by wavelengths) you need to know which end of the spectrum is which. Thus, from the Kodak graph for their 'Wratten 85' series of amber colour temperature conversion filters (above) it can be seen that an 85B eliminates more 400 nm or m$\mu$ (violet) light than an 85 or 85C. An 85N6 can be seen as effective in the 424–450 nm region and so would be useful for high altitude mountain and snow cinematography where the predominance of blue may be a problem.

# Exposure

Filmstock will only record an image within a limited range of light intensity. Above and below this range detail which may be seen by the human eye is not recorded on film. The aim in assessing exposure for photographic purposes is to set this band of recordable light within the band of most useful details in the subject to be recorded. If the exposure is too great the shadow detail will be recorded at the expense of the highlights, and if too little, only that which is most intensely illuminated is recorded.

## Latitude

A cinematographer who uses a particular filmstock should make frequent tests to see for himself how much his exposure may be set either side of the ideal and still produce a usable exposure. In assessing exposure latitude he should look particularly at the density and colour of the shadows and the detail in the highlights.

The amount of underexposure which is acceptable is closely related to the subject brightness range. The lower the contrast, the greater the exposure latitude. Grossly underexposed scenes cannot be intercut with those normally exposed without their deficiencies becoming noticeable. It might be preferable to underexpose every shot of an entire sequence to hide away the problems as an overall effect rather than risk patchiness.

Certain negative colour stocks give a quite acceptable image when at least 1–1½ stops over- or underexposed. Even when 2–3 stops incorrect all may not be lost, although there would be noticeable deficiencies in either the shadows or the highlights.

Reversal stocks, particularly very fast ones, have little latitude by comparison and require exposure within ⅓–½ of a stop of 'correct'. At one stop incorrect, the image is virtually unusable by professional standards.

Filmstocks with poor latitude characteristics apart from demanding a very much higher standard of exposure assessment, show less detail in the shadows even if correctly exposed.

Having made tests, the cinematographer should mount individual frames in slide holders and project them as stills. In this manner he may study the effects at leisure, draw his own conclusions and determine the limitations which he must apply within his particular field of endeavour.

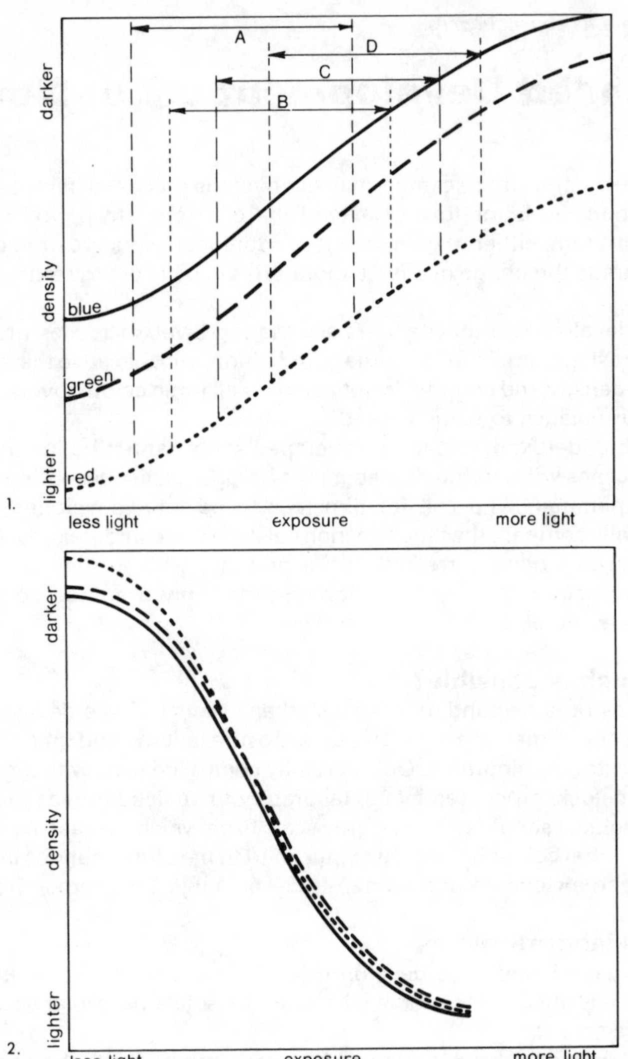


EXPOSURE

1. **Colour negative**
Typical relative log curves of a normally developed colour negative filmstock (ECN II). Note how normal exposure compares with the negative one and two stops underexposed, and one stop overexposed. In all cases the exposure can be corrected in the printing stage. A, 2 stops under; B, 1 stop under; C, normal; D, 1 stop over.

2. **Colour reversal**
Typical relative log curves of a normally developed colour reversal filmstock (7250). Note how underexposed red tends to become more dense (darker) compared with the blues. If this is compensated for in the printing stage, the balance of colours in the correctly exposed parts of the negative will be upset.

# Extended Development ... or Not?

One of the controls that a cameraman has over the process of film-making is to call upon the laboratory to extend the development period or force develop his film, either to gain apparent additional exposure or to deliberately degrade the image quality. Laboratories usually charge extra for this facility.

Force-development inevitably raises the fog level, increases grain size and contrast (gamma), reduces image definition, and, because the increase in image density and contrast is not equal for all three colour layers, upsets the colour balance to some extent.

As with underexposure, force-developed shots cannot be intercut with normal scenes without the degradation of image quality becoming noticeable. Cameramen who call for filmstock to be force-developed should occasionally compare the results of normal developed underexposed negative which has been corrected in the printing, with a print of a force-developed negative. They may well find that only a limited amount of forcing is acceptable.

### How much is possible?

Some filmstocks respond more readily than others to force-development. Equally, some filmstocks have greater exposure latitude and therefore less need for extra development. Only carefully controlled tests with the appropriate filmstock, processed by the laboratory to be used proves the point. The test subject should contain a grey scale from which to measure density and the neutral colour balance, a colour chart to measure changes in colour balance and a close-up of a human face—no subject is more important.

### Tell the laboratory

When contemplating force-development, the entire roll must be underexposed, as it is impossible for the laboratory to isolate particular scenes for special treatment.

Films to be force-developed must be clearly marked with the necessary instructions on the label of every can as well as on the negative report sheets which accompany the film to the laboratory.

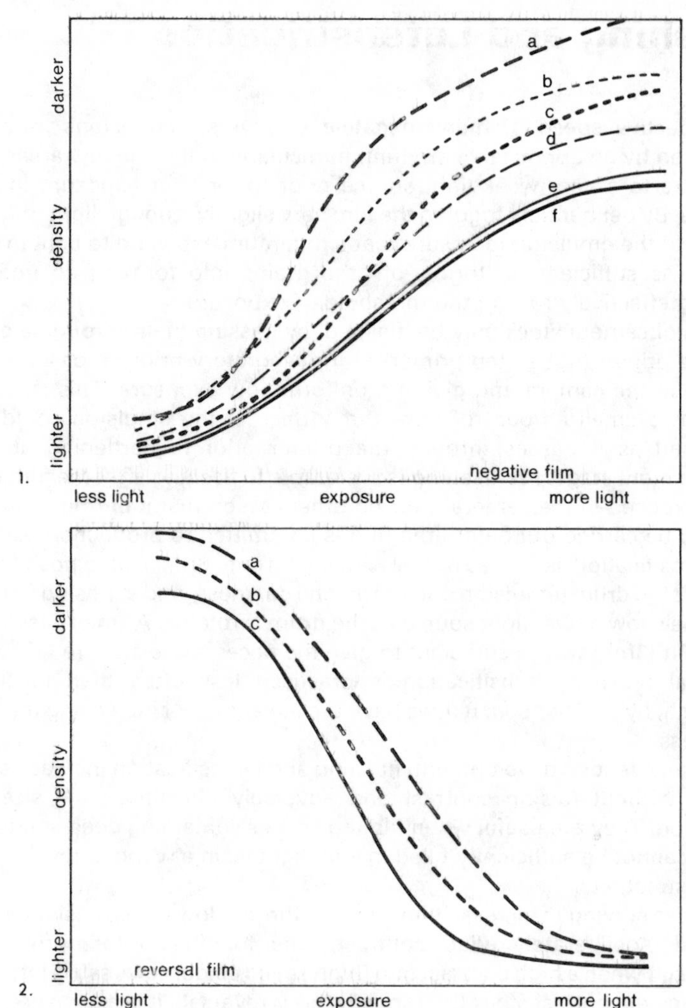


EXTENDED DEVELOPMENT

1. The effect of extended development, blue and red records only, of Kodak ECN II. The effect on the green record is similar to the blue. Note how the contrast and density of the red record increases by comparison to the blue. For this reason the shadow areas in prints made from forced negatives tend to look blue-green when the flesh tones are graded (timed) to look normal. a, red forced 2 stops; b, blue forced 2 stops; c, red forced 1 stop; d, blue forced 1 stop; e, blue normal development; f, red normal development.
2. The effect of extended development, blue record only, of Ektachrome reversal Type 7240. Note that contrast increases with forcing, that the maximum density decreases and that there is little change in the highlights. a, blue normal development; b, blue forced 1 stop; c, blue forced 2 stops.

*A means of increasing shadow detail.*

# Flashing and Latensification

The effective speed of many negative or reversal emulsions may be increased by an appreciable amount, particularly in the shadow areas, by exposure to a very weak light source prior to, or after exposure in the camera. By deliberately fogging the film very slightly, enough light may be put on to the emulsion to ensure that a minimum exposure to light in the camera is sufficient to 'trigger-off' the grains into forming an image, thereby effectively raising the threshold of exposure.

Rolls of camera stock may be 'flashed' by passing them through a continuous printer (not a step printer) at the laboratory prior to, or after exposure in the camera and giving a uniform, low exposure. This method involves a small amount of risk—the surface of the emulsion could be scratched as it passes through the printer prior to hardening at the development stage. Pre-flashing is preferable to flashing after the film has been exposed in the camera if for no other reason that it permits control tests to be carried out before the film is committed to production use.

'Latensification' is carried out by winding an entire roll of unexposed film onto a large drum (emulsion out and in total darkness) and exposing it to an extremely low power light source as the drum is rotated. A small flashlight bulb 10ft (3m) away is sufficient to give the necessary exposure 'lift'. For practical reasons latensification is practised less often than flashing although, by comparison, it does have the advantage of increasing the fog level less.

Unlike extended development, flashing and latensification increase sensitivity without raising contrast and adversely affecting grain size or definition. They are useful when filming scenes containing deep shadows which cannot be sufficiently filled in with light ie. in a wood, a ravine or a shady street, etc.

Besides serving to increase film speed in the shadow areas, flashing may be used to deliberately reduce contrast, either for effect or to improve the final result when a high contrast film (high speed colour reversal) is intercut with a lower contrast stock (low speed colour reversal). It is also used with negative stock that is to be given extended development, to offset the increase in contrast that would otherwise result.

If flashing or latensification are done with a coloured light source either process adds an overall colour cast to the shadow areas of a scene without affecting the highlights (and faces) which would not be the case if a coloured filter were used on the camera or in the printing.

When a roll of film is flashed or latensified extreme care must be taken to ensure even exposure in the camera, if the subsequent grading (timing) is not to upset the balance between the two exposure levels.

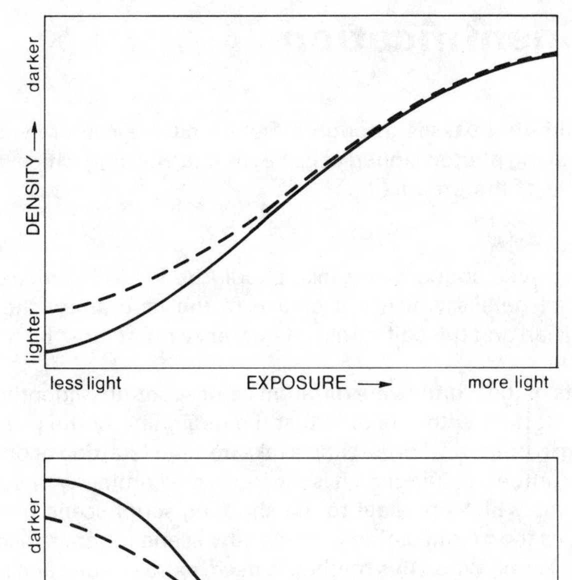


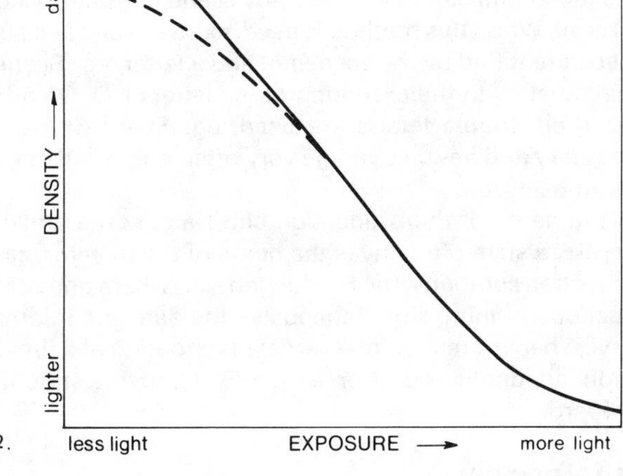


FLASHING AND LATENSIFICATION

The effect of flash exposures on the red records of negative and reversal stocks. To maintain colour balance, the increase of fog level on the green and blue records must be made substantially the same.
1. Negative ECN II. Solid line normal exposure curve, broken line, as a result of flashing. Note that density is built up in the areas of less exposure (shadows) to a greater extent than in the brighter areas, causing a decrease in contrast.
2. Reversal (Ektachrome). Solid line normal exposure curve, broken line, as a result of flashing. Note that the density is reduced to a greater extent in the shadow areas than in the highlights, causing a decrease in contrast.

# Visual Identification

Every roll of film that passes through a motion picture camera should be identified by having photographed at its beginning wording which, at least, states the name of the production.

## The slate

When shooting sync sound every take should be identified visually and verbally with all details which are of use to the laboratory, the sound transfer technician and the editor; also a reference point must be made for synchronisation.

In some parts of the world every different camera set-up is identified by a slate number (starting with one or 100 at the beginning of the production and continuing in numerical order) and a take number (starting at one every time the slate number is changed). It is the job of the continuity or script girl to provide sheets which co-relate to the shooting script scene numbers.

In other places the identification is directly by scene numbers according to the shooting script. When this method is used, extra set-ups and subsidiary inserts must be identified by the scene no. and a letter, eg: Scene 88A, and when the alphabet, with the exception of the letters I, O, Q and Z has been exhausted, then double letters are used, eg: Scene 88AA. When shooting documentary and news coverage very often only a roll number is shot at the head of each roll.

In addition to the name of the production and slate, scene and/or take identifying numbers, a slate often gives the name of the director, cameraman and the production company, the Production No. (where one company has many productions running simultaneously), the date and information for the laboratory. When more than one camera is shooting at a time, each camera must shoot an identifying letter, as A or B etc, before shooting the slate or clapperboard.

## Laboratory information

This may consist of grading (timing) instructions in relation to the overall effect required. Normal exposure being marked DAY/INT(erior) or DAY/EXT(erior).

Film deliberately shot 'low key' for night effect and which should be printed accordingly is marked as NIGHT/INT or EXT or DAY FOR NIGHT. In France *day for night* is described as *la nuit Americaine*.

Greyscale and colour scale cards are also sometimes attached to the slate for benefit of the laboratory. This is particularly useful when graded (timed) overnight rushes (dailies) are made, when it may serve as an initial guide for the laboratory.

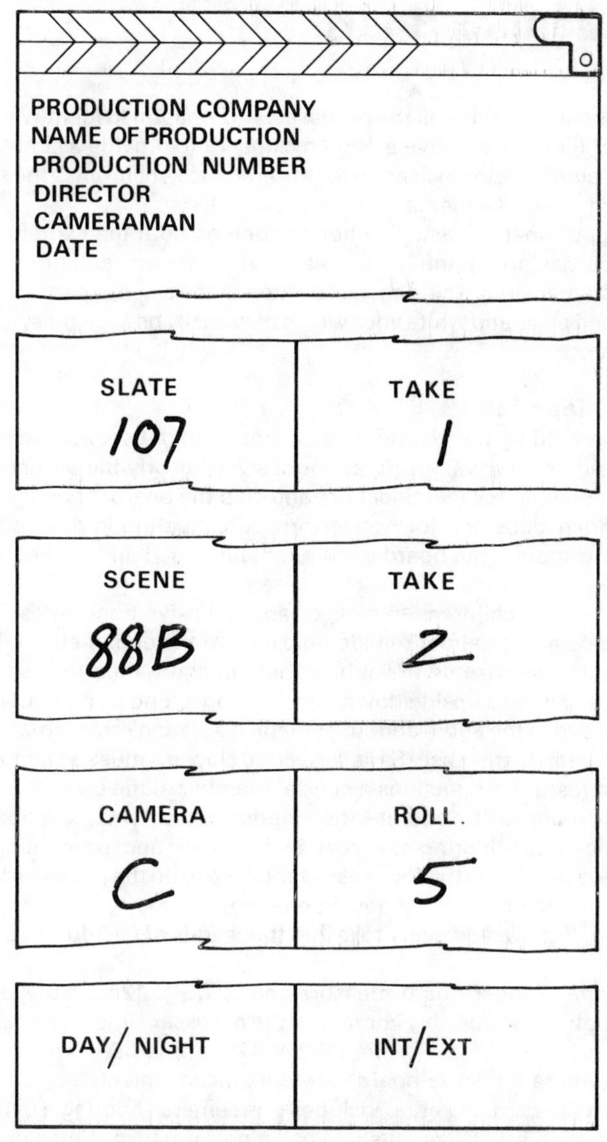


TYPICAL SLATE-BOARD INFORMATION

This shows production identification information, sequential slate numbers, numbering according to scripted scenes, documentary numbering and laboratory information.

# The Clapper Board

A clapper board contains all the necessary information to identify a particular piece of film; it may have a hinged clapstick to provide a simultaneous visual and audible reference point for the editor to synchronise the separate picture and sound elements.

The clapper board is usually shot as soon as both the camera and the sound recorder are running to speed. With modern equipment this is almost instantaneous. The clapstick portion of the clapper board is marked with hatched black and white lines which may easily be seen for synchronising purposes.

## Using a clapstick

The person holding the clapper board should hold it very steadily, otherwise blurred images will result, announce very clearly the pertinent information for soundtrack identification, and clap the board. Loud shouts and heavy handed claps are not necessary when shooting in quiet surroundings. After clapping, the board should remain closed and in shot for a few moments.

When shooting children, animals or apprehensive non-professionals, it may be preferable to shoot the identification at the end of each take so as not to frighten or unsettle the subject beforehand. When this is done the clapper board is held upside down and the words 'End clap' or 'Clapper on end' are added to the audio announcement. If a slating error is made, it may call for a repeat. In this case 'Sync to second clapper' must be announced.

On most feature productions a special member of the crew, the 'clapper loader' is employed to operate the clapper board. This is economically advantageous considering the cost of filmstock and processing which would be wasted if say, the focus assistant were to do the job and spend five seconds on each take settling down before pulling focus. This would mean $7\frac{1}{2}$ft ($2\frac{1}{2}$m) of film wasted every take and thousands of feet during a production.

The normal clapper board measures about 15 × 12in (380 × 300mm), large enough to comfortably contain all the necessary information. A smaller board, usually about 6 × 5in (150 × 125mm) is used for inserts and close-ups where the large board would be inconvenient.

When a multi-camera set-up includes one camera shooting a particularly wide angle shot, say, a live on-stage ballet performance, it may be necessary to use a 'king-size' clapper board visible to the distant camera. It is essential to fill the screen with the clapper board even if it involves zooming. Camera crews who provide the editor with wide angle clapper board shots should be made to attempt the syncing themselves!

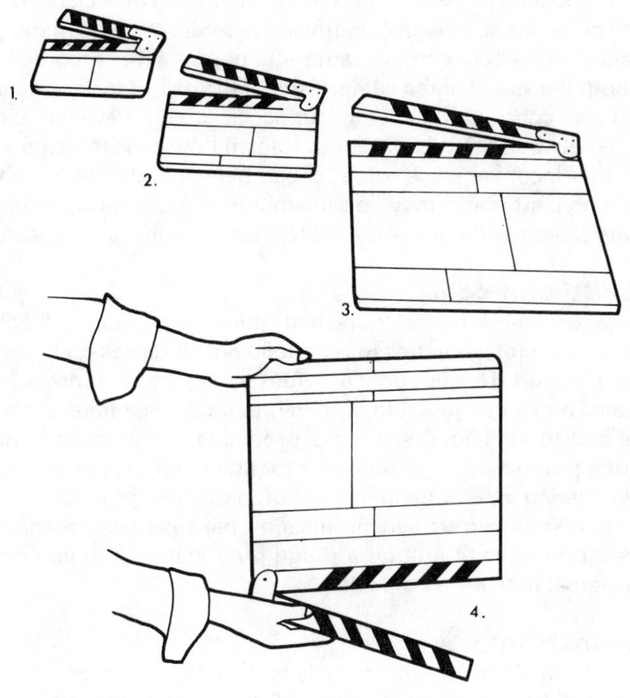


THE CLAPPER BOARD

Various sizes of clapper board are made for different types of shot (1, 2, 3). The smallest size takes account of close-up shots, where a standard size board is inconvenient to handle. The largest board is for use with distant camera placement. For clapper-on-end shots, where the sync point and references appear at the end of the shot instead of the beginning, the board should be held upside down (4).

*Exotic alternatives to the clapper board.*

# Sync Marker Light and Time Code Systems

Many cameras and sound recorders incorporate a sync marker system which, provided the sound recorder is started before the camera, automatically causes one or two film frames to be fogged and the tone to be superimposed on the sound track at the beginning of each take. In most cases the connection between camera and recorder for sync marking is by a lightweight cable between the camera battery and the recorder.

When both the speed of the camera and of the recorder are regulated by quartz crystal control, it becomes unnecessary to have an electrical connection between the two except to transmit the sync marking system. To save the inconvenience of a cable between the two units, the synchronising start marks may be transmitted by radio, or both units may incorporate a time-code generator which achieves the same results.

## SMPTE/EBU time code

Time codes are an electronic equivalent of film edge numbers but using magnetic recording techniques to record 80 on/off pulses on every frame of picture or sound. This not only provides an electronic address for each frame, it also notes the direction of travel; uniquely identifies every frame of picture and sound shot on an entire production; distinguishes material shot for one production from that shot for any other production; enables automatic finding and indentification of particular frames or scenes; automatically finds correct synchronisation between picture and sound; and makes it possible to edit on a rough print and then to automatically edit the original master to conform.

## Longitudinal time code

Longitudinal SMPTE/EBU time code is laid lengthways along the film or sound track. Each frame consists of 80 equally spaced on/off pulses irrespective of whether the film is travelling at 24 or 25fps. Any of the pulses may be switched halfway through their period to become 'bits' of information which are subsequently decoded to read the frame number, the minute, the second, the hour and the direction the film is travelling (sync word). In addition, 'unassigned bits' or 'user bits' may be used for information such as reel numbers, recording date, production identification etc. provided it is recorded simultaneously with the time code.

## Vertical interval time code

VITC SMPTE/EBU time code is used with video recorders and consists of 90 bits recorded during the vertical blanking interval of each field/frame. VITC is the preferred system for video because it is less prone to losing readability and can be read at faster and slower speeds. When used with video, time codes also carry such information as field identification, colour frame identification and 'drop frame' information to compensate for the fact that NTSC colour signals have an actual frequency of 29.97fps.

## TIME CODE DATA STRUCTURE

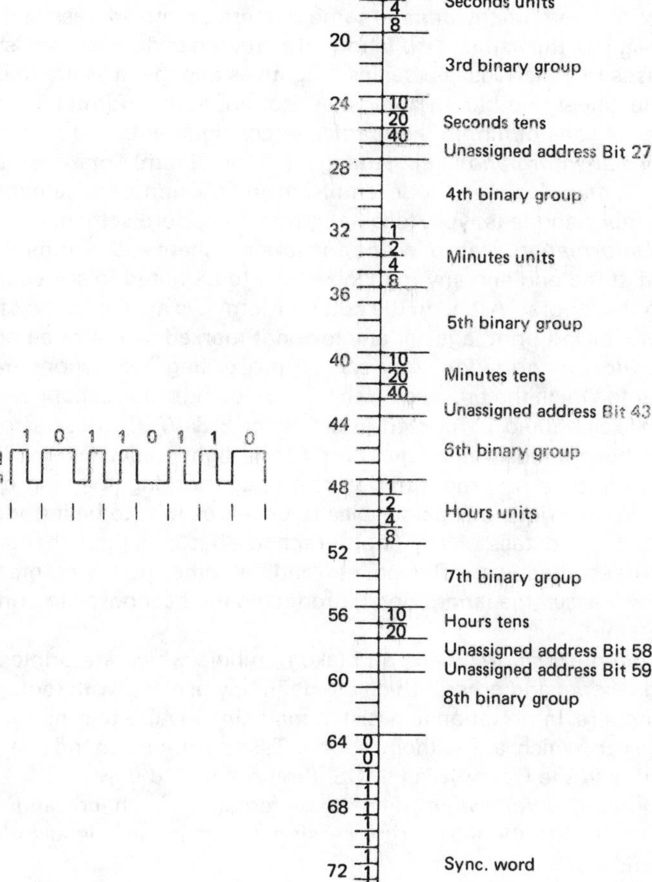

# Camera Report Sheets

All camera original film shot to a script must be accompanied by a document which details the contents and acts as an official order for processing and printing.

At the end of a day's shooting the camera assistant should check with the continuity (script) girl or production assistant and the sound recordist that the information is all correct. Copies of the sheets should be sent to the laboratory, the production office, the accounts department, cutting room and one be retained for himself.

Camera report sheets usually detail the following facts:

*Camera information*. The name of the production company, name and number (eg: TV series) of the production, the names of the director, the director of photography and the camera operator, the address of the location or studio, the name of the laboratory ordered to do the processing, the addresses to which rushes (dailies) negatives and invoices are to be sent, the date, sheet number and any sheet continuation information, and the number of cans of film in that particular consignment.

*Technical information*. Film gauge (16 or 35mm), make or type of filmstock, manufacturer's code, emulsion and roll numbers, camera model and number and lens type, focal length and aperture setting.

*Shot information*. Slate or scene and take numbers and lengths. Any take marked at the end and any marking errors to be noted to save confusion both in the laboratory and in the cutting room. It is also important to mark 'no slate' or 'no print' against any take not marked or not to be printed.

*Laboratory information*. Any special processing instructions (ie: ASA number to which the processing should be forced), instructions regarding which takes should be printed in colour or B & W, if the selected takes should be graded (timed) or printed 'one light'; advice regarding the desired effect to be produced (ie: day, early morning, evening or night, interior or exterior); any colour bias to be left as it is, to be introduced or corrected for; details of any photographed effects as light changes—fire flicker, deep shadows, diffusion, etc. and any other pertinent information which will assist the laboratory in producing the best possible print at the first attempt.

*Editing information*. Scene and take numbers which are printed, those held in reserve and others which are definitely useless, with footages and can numbers. Information to assist in matching picture to sync sound and notes as to which are without sound. Takes without sound are usually marked with the Goldwynian 'MOS' ('mit out sound').

*Accounting information*. Filmstock footages in hand and drawn, exposed, sent to the labs for processing, wastage, and details of usable short ends.

CUTTING ROOMS COPY  №. 49782

| CONTINUED FROM SHEET No. | SHEET NUMBER ONE | CONTINUED ON SHEET No. |
|---|---|---|

THE SHEET NUMBERS MUST BE QUOTED ON ALL DELIVERY NOTES, INVOICES AND OTHER COMMUNICATIONS RELATING THERETO

PRODUCING COMPANY MOODY PRODS. LTD  STUDIOS OR LOCATION BASINGSTOKE

PRODUCTION 'EAT MORE JELLY'  PRODUCTION No. 77/36

DIRECTOR G. FROMAGE  CAMERAMAN W. FOGG  DATE 6-11-77

STATE IF COLOUR OR B & W: COLOUR

# PICTURE NEGATIVE REPORT

ORDER TO BEST LABS. LTD LABORATORIES

| STOCK AND CODE No. | LABORATORY INSTRUCTIONS RE INVOICING, DELIVERY, ETC. | CAMERA AND NUMBER |
|---|---|---|
| E/C 5247 | RUSHES TO BE DELIVERED FIRST | PFX 77 |
| **EMULSION AND ROLL No.** | | **CAMERA OPERATOR** |
| 262-17 | VAN TO 242 WARDOUR STREET. | C. FARNSBARNS |

| MAG. No. | LENGTH LOADED | SLATE No. | TAKE No. | COUNTER READING | TAKE LENGTH | 'P' FOR PRINT B & W | 'P' FOR PRINT COL'R | LENS F/L & STOP | ESSENTIAL INFORMATION ★SEE REQUIREMENTS BELOW | CAN No. |
|---|---|---|---|---|---|---|---|---|---|---|
| 1 | 280 S/E | 1 | 1 | 40 | 40 | — | | | ALL INT. DAY | |
| | | | 2/5 | 220 | 180 | | P | | | (1) |
| | | | 6/7 | 270 | 50 | — | | | 10ft. waste. | |
| 2 | 400 | 1 | 8/9 | 80 | 80 | | P | | | |
| | | 2 | 1/3 | 120 | 40 | — | | | BOARD ON END | (2) |
| | | | 4 | 400 | 280 | | P | | PRINT TO END (NO WASTE) | |
| 3 | 400 | 3 | 1 | 60 | 60 | — | | | | |
| 6 | 400 | 21 | 1 | 80 | 80 | — | | | | |
| | | | 2 | 200 | 120 | | P | | — MUTE | |
| | | 22 | 1 | 240 | 40 | | P | | | |

FOR OFFICE USE ONLY  TOTAL CANS

| TOTAL EXPOSED | 2120 | TOTAL EXPOSED | 2120 | TOTAL PRINTED | TOTAL FOOTAGE PREVIOUSLY DRAWN | 280 |
|---|---|---|---|---|---|---|
| SHORT ENDS | 160 | HELD OR NOT SENT | — | B & W — | FOOTAGE DRAWN TODAY | 2000 |
| WASTE | | TOTAL DEVELOPED | 2120 | COLOUR 1240 | PREVIOUSLY EXPOSED | |
| FOOTAGE LOADED | | | | | EXPOSED TODAY | 2120 |

SIGNED :

★COLOUR DESCRIPTION OF SCENE. FILTER AND/OR DIFFUSION USED. DAY, NIGHT OR OTHER EFFECTS, DAYLIGHT, ARCS, INKIES OR MIXED LIGHTING. INTERIOR/EXTERIOR A.M., P.M.

S.L.P. 1087

CAMERA REPORT SHEETS

Part of the typical negative report for a TV commercial production.

# Dope Sheets

Film shot 'off the cuff' newsreel fashion must have a more journalistic type of information sheet to accompany it than the camera report which goes with scripted film.

While it is generally accepted that in newsreel shooting conditions it is not possible to number every scene and take, it is nevertheless vitally important to visually number each roll even if by the simple expedient of holding up the requisite number of fingers.

'Hot' news film very often comes straight off the processing machine to be screened or viewed and edited and if for TV, transmitted without a print being made. Where time permits, as when shooting a documentary record of an event, a print is made off the entire camera original film (print all) without regard to scenes or shots or printing selected takes only for reasons of economy.

## Shot list, etc.

In these circumstances, the cameraman should write a comprehensive shot list to explain the coverage in detail to the editorial department. The details of the event photographed, the names of people and places, the type of shot (close-up, medium and long shot) and any other relevant information, must be clearly stated if the film is to be usable. A news editor sitting in a viewing theatre cannot be expected to imagine what the cameraman had in mind when he shot certain scenes. If the cameraman wishes to have his material used to its full worth, he must supply comprehensive captions just as a stills photographer would.

After its initial use, important newsreel material is stored in a library for posterity where, again, every detail should be documented on paper.

A newsreel cameraman's dope sheet usually contains the following information in addition to a roll by roll shot by shot description of the coverage: the name of the newsreel company and bureau location, the date, subject, location, names of the cameraman, sound recordist and reporter, make and name of the filmstock and manufacturer's code number, film speed to which the film is to be processed, the footage exposed and the names of any opposition companies also covering the event.

The required number of completed dope sheet copies should be attached to the film cans with adhesive tape when they are sent to the laboratory. A copy should be retained by the cameraman so that he may telephone the details to his office prior to the film arriving, to keep for his records and to use as an *aide mémoire* when filling out his expenses sheet.

## WORLD NEWS LTD.

FORM 260

**CAMERAMAN'S REPORT**

CAMERA CREW TED FISHER | SOUND CRUW DAVID McGUIRE

ASSISTANT R HARD KNIGHT | LIGHTING CREW JIM BALDWIN

DATE & LOCATION JUNE 25th 1973 : STROUD, GLOSTER | COMMENCEMENT 09·00 hrs | COMPLETION 17·30 hrs

ASSIGNMENT SALE OF TREASURES OF THE LATE LORD MILL | JOB NO. FOR GENERAL RELEASE

| FOOTAGE | WASTE | LIGHT CONDITIONS | SOUND |
|---|---|---|---|
| 680' E/C 5247 | 20' | ART | WILD ATMOSPHERE SYNC. AUCTION |
| **CAMERA** | **SOUND EQUIPMENT** | **LIGHT EQUIPMENT** | **CAR NO. MILEAGE** |
| BL ARRI (SFS.) | NAGRA IV | MAINS QUARTZ | PCD 464G 250M. |

REMARKS

ROLL 1 FORCE DEV. ONE STOP PLEASE
FOR PRICES & BUYERS SEE ATTACHED CATALOGUE.

An accurate description of each individual scene including names of persons figuring therein is necessary.
In group shots always give names from left to right, making certain they are spelt correctly.

WRITE CLEARLY TO ALLOW FOR PHOTOSTAT COPIES TO BE MADE

SCENES:

ROLL 1 120' GV. INT SHAM HALL ① ENTRANCE HALL SHOWING FAMOUS STAIRWAY ② DITTO, ③ LIBRARY. ④ STUDY
ROLL 2 280' CU ITEM 106 IN CATALOGUE, DITTO 108, 117, 134, 135 & 137.
GV AUCTION (in dining room).
3 Shots same
CU AUCTIONEER
ITEM 106 AUCTIONED.
CUTAWAYS BIDDING.
ITEM 108 AUCTIONED
Sir JAMES MILCHIN BIDS.
ROLL 3 280' ITEM 117 AUCTIONED
CUTAWAYS INCLUDE POP STAR ROCK HILTON
ITEMS 134, 135 & 137 AUCTIONED
CUTAWAYS
G.V. AUCTION
G.V. EXTERIOR.

NB ITEMS 134 & 135 DID NOT REACH RESERVE & WERE WITHDRAWN. TOP PRICE IN SALE ITEM 137 by ROCK HILTON (SEE CATALOGUE).

ALSO COVERED BY: BBC / ITN / VISNEWS / CBS / ABC.

*Sound recordists are human too.*

# Shooting Dialogue

Post-synchronising sound is a time consuming and expensive process, which can entail recalling artists who may by then be engaged on other productions. Sometimes an artist can repeat lines in a state of quiet after a noisy take, while the timings and nuances are fresh in the mind. A TV viewfinder playback may be useful in this respect. However, if possible, post-synchronising should be avoided.

**Cameraman—soundman co-operation**

For the sake of the production cost it behoves the cameraman to work as closely as possible with the sound recordist to achieve usable sound which does not entail a post-synchronisation session.

Co-operation is a two way affair. It is no use for the recordist to place an omnidirectional microphone on a boom in the middle of a set with a take it or leave it attitude. He must be prepared and equipped to use highly directional line microphones set at angles which minimise extraneous noise. He may also need very small personal microphones that can be concealed in an artist's clothing. These feed into small radio transmitters or miniature recorders worn by the artist. Microphones may be hidden in the set, hung overhead, placed low, carried by one artist to pick up another, and so on.

Micro-miniature microphones may be placed in an artist's button holes etc. and are so small (about the size of a match head) that it does not matter if they are seen by the camera.

The cameraman, by his lighting, can conceal microphones in created shadows, can place objects to hide them and may adjust the camera angle to help.

It may be necessary to use a quieter camera, to cover a noisy camera with blankets or a barney (16mm cameras especially, are notoriously noisy), use a lens which does not trumpet the camera noise, if this is a problem, shoot through a glass fronted blimp, hang drapes about the set to cover acoustically reflective and resonating surfaces, place the generator further away downwind or on the other side of a substantial building, or arrange baffles between the noise source and the microphones.

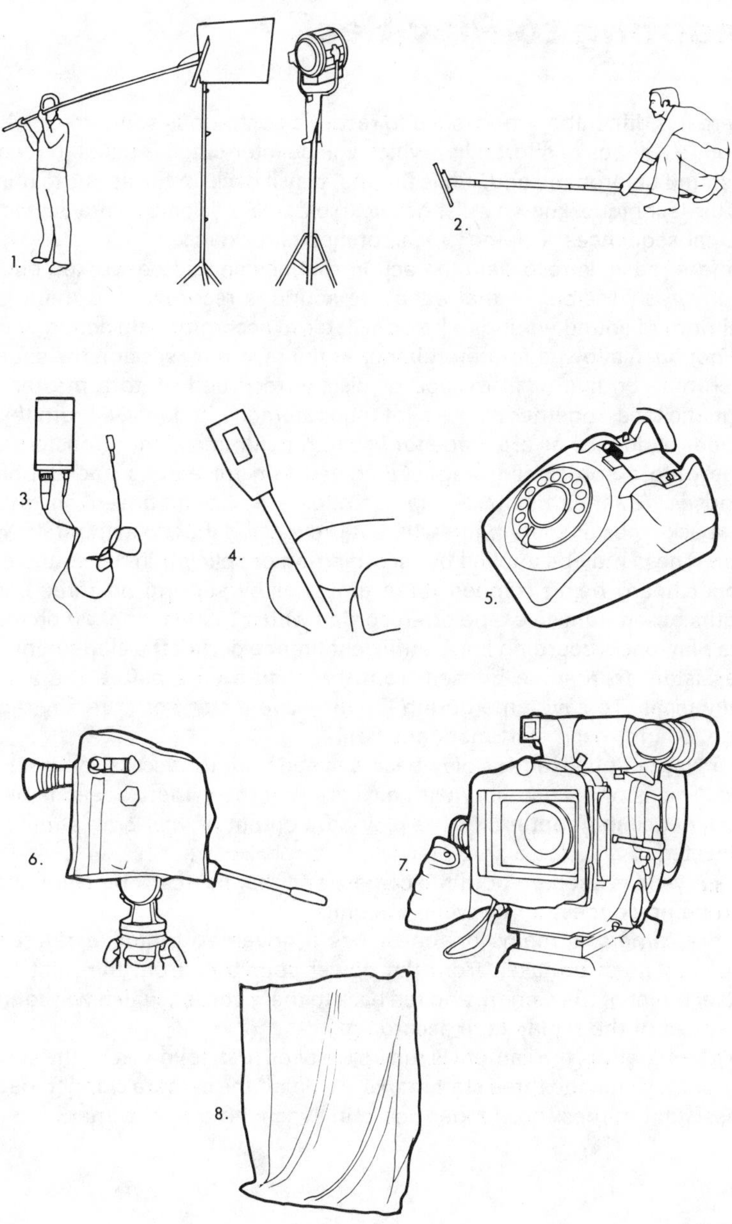

SHOOTING DIALOGUE

1. Microphone held behind created shadow; 2. Microphone held low; 3. Radio microphone; 4. Miniature microphone; 5. Hidden microphone; 6. Camera enclosed by sound barney; 7. Lens behind glass; 8. Curtain hung in front of sound reflecting surface.

# Shooting to Play-Back

Where it is difficult or impossible to record a continuous soundtrack while shooting a series of short takes which will be inter-cut, it is usual to record the sound beforehand and, while filming, play it back for the artists to mime to. This technique, known as 'shooting to play-back' is particularly suited to musical sequences, whether vocal, orchestral or dance.

Before music is recorded, the action routines should be worked out as carefully as possible so that adequate sound is recorded and there is a minimum of sound editing to be done later to accommodate action which had not been allowed for beforehand. At the recording session the sound, which may be live or from tape or disc, is recorded on to a master ¼in magnetic tape, together with a pilot tone reference frequency. From this a number of duplicates are made for use during filming. Other transfers are made onto a sprocketed magnetic coated film for editing and dubbing purposes. A series of three claps or blips are superimposed on to the play-back tapes at every point where it is expected that an action-take will begin. These may be applied by recording-on or splicing-in the sound of a clapper board being clapped three times, or by sticking on three short lengths of self-adhesive tape pre-recorded with a 1000Hz tone. When making a play-back recording leave sufficient time to permit the clapper/loader or assistant to remove himself from the picture area before the action commences. To save time during filming have a number of replay tapes made, one for each start mark position.

If a Nagra III is used for play-back a speed variator accessory must be used. This is only necessary with a Nagra IV if the camera speed is more than 4 per cent off optimum. The play-back output of the Nagra should be connected to a suitable amplifier and loudspeaker system.

If a crystal is used on both the camera and the play-back machine there need be no connection between the units.

If the camera is mains driven or has a governed motor, a reference frequency must be taken from the power supply or from the pilot tone (pulse) outlet of the camera and fed back to the recorder, which will control the speed of the replay tape accordingly.

When shooting, the camera is run up to speed first, followed by the sound play-back. When the three start marks are heard the camera clapper/loader or assistant mimes the clapper board in sync with the third mark.

SHOOTING TO PLAY-BACK

Nagra equipment for play-back to be used in conjunction with a quartz crystal speed-controlled camera.

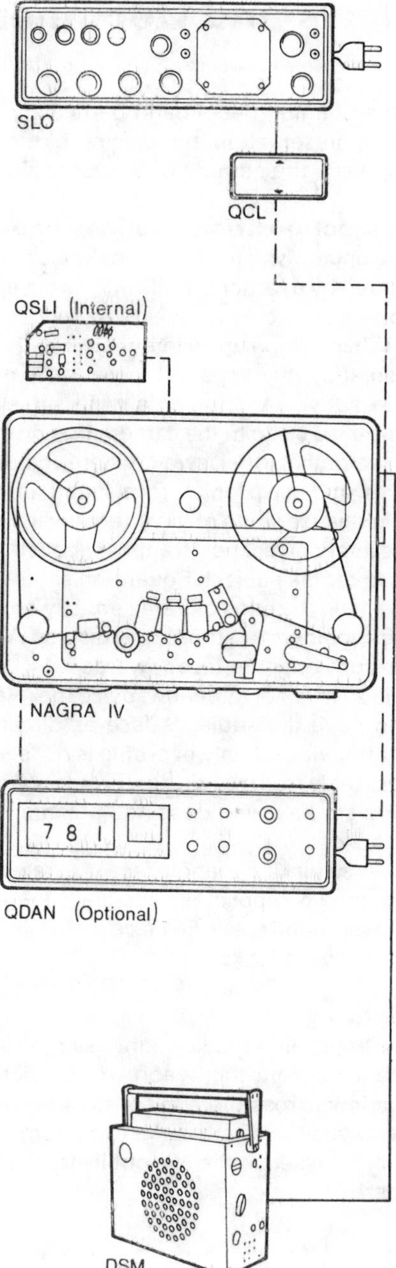


**SLO**
Synchroniser to control play-back speed in order to phase lock the speed control signal on the pilot track with internal, external or mains reference frequency, with cathode ray tube display and plug-in SLQ quartz-controlled generator; built-in recorder power supply.

**QCL**
Adapter for connecting the SLO to the recorder. SLO and QCL must be used with Nagra III and may be used as an alternative to QSLI which must be fitted internally within the Nagra IV.

**QSLI**
Synchroniser to control play-back speed in order to phase lock the speed control signal on the pilot track with an internal or external reference frequency.

**NAGRA IV**
Used as a play-off machine for a tape which has been pre-recorded with both sound and quartz crystal pilot tracks.

**QDAN**
Decoder with digital display for the location of recorded sequences and preselection for automatic stop.

**DSM**
Amplifier/loudspeaker unit.

# News and Documentary Interviews

When filming people and particularly interviews for TV, documentary or news presentation, the placing of the camera relative to the subject may influence the attitude of viewer to the subject.

## Subject position and lens choice

Looking down on a person belittles him, looking up boosts his importance. Filmed by a camera level with his eyes, both the subject and the viewer are, so to speak, on the level.

When setting up an interview, relative sizes of the subject and interviewer also affect the apparent importance of the subject.

In a 'two shot', using a wide angle lens the interviewer should not be placed so close to the camera that he overpowers the subject. It is better to place both further away, where the perspective 'evens out' their relative sizes, and use a longer focal length of lens to gain the image size you want.

In reverse shots of the interviewer alone, filmed after the subject has left, the interviewer should appear about the same size in the picture area as that used for the subject. Square-on shots of a person with his or her shoulders touching the edge of frame equally on either side invariably look ugly. Turn the shoulders to a three-quarter view. When looking at the camera let him or her turn only the face towards it.

In a two-shot interview, place the interviewer with his back to the camera and have the subject's face as directly towards the lens as possible. A subject viewed only in profile is not seen at his best, for few people look as agreeable from the side as they do full face. He will not be convincing either, for a speaker who does not look the audience in the eye will convince no one.

When placing a single face in a frame, a very good rule invariably producing good composition, is to place the subject's nose on the centre marking of the ground glass. Full face or profile, head and shoulders or face only, the shot always looks right.

## Lighting

Try to position the key light as high as possible without losing the eyes in shadows from the eyebrows. Do not let the keylight cast an ugly nose shadow across the subject's cheek. Avoid two nose shadows. Beware of nasty shadows across the eyes from spectacle frames. Try to get pinpoint light reflected in the eye pupils from a filler light. Use as little extra light as possible.

SHOOTING INTERVIEWS

1. Looking up to a person boosts their importance; 2. Looking down belittles them; 3. Wide angle lenses make the subject insignificant; 4. Long lenses make people equal; 5. Square-on shoulder shots look ugly; 6. Slightly side-on view makes a person look slimmer; 7, 8. To put the subject's nose absolutely in the middle of the screen is usually correct.

# Prompting Devices and Idiot Boards

When filming interviews or statements for documentary films, it not infrequently happens that the subject either becomes speechless when confronted with a camera or the statement is too long and full of facts and figures to be memorised completely, or the subject may be required to say a few words in a language in which he or she is not familiar. To avoid a frustrating shooting session in such cases, some aid is needed.

## Prompting systems

A number of sophisticated prompting systems are available, on a contract basis, complete with experienced personnel. They fit their devices on to any type of camera, write out the script on the display sheet, and operate the device keeping the script in step with the speaker as he reads.

The device may use a paper roll, or a TV monitor set close to the camera lens and visible to the speaker via a partial reflecting mirror set at an angle in front of the camera lens so that his eyes look directly into the lens, and so at the viewer.

## Improvised systems

Closed circuit TV may be used to make a simple prompting device. A CCTV camera can scan a typewritten script, coupled to a TV monitor set immediately below or above the film camera lens. Even without a partial mirror this system is highly successful and is used by many TV stations for news broadcasts.

## In an emergency

It is when shooting on location and least prepared and equipped that you are most likely to encounter interviewees with partial or total amnesia or dyslexia. A few key words written on large cards or sheets of paper with a felt tipped marker pen can produce wonders in jogging the memory. Subjects who cannot speak spontaneously, and are also dyslexic need rehearsal.

If a hole is cut in the centre of the card it is possible to keep all the writing close to the lens. If the lettering is too far from the lens the speaker may tend to look shifty-eyed and less than honest. If the reader would normally require spectacles to read from such a distance but is not doing so while being filmed, move the camera closer or use larger lettering. Remember that lettering looks smaller from the subject's position than when it is being written.

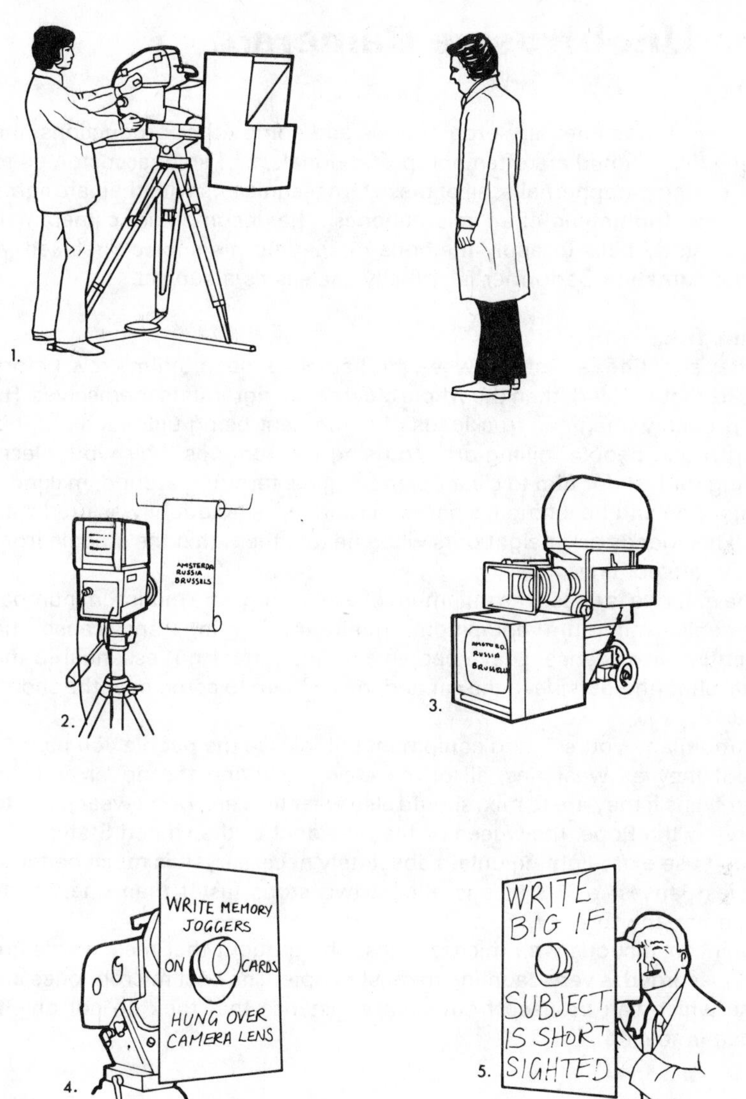


PROMPTING DEVICES

1. Autocue system with script reflected in semi-silvered mirror placed in front of camera lens; 2, 3. Cueing system using CCTV camera looking at script type on roll of paper and TV set placed below film camera lens; 4, 5. Idiot boards.

*Cinema-verite technique.*

# The Unobtrusive Camera

In many TV documentary programmes and some other productions, the people to be filmed are often non-professionals, and so unaccustomed to film-making paraphernalia, all of these film technicians with their arc lights and, most frightening of all, microphones. The documentary cameraman should study hard to apply methods of keeping his subjects relaxed. A nervous amateur performer is virtually useless as a subject.

**Least fuss**

If, after shooting, an interviewee who has never seen a film crew before says 'Is that all it is?', then the whole crew can congratulate themselves. He had probably imagined truckloads of equipment being unloaded into his living room, people milling around using his home as their work place, leaving rubbish for him to clear up, moving his furniture around, making a commotion and upsetting his peaceful routine, as no doubt was predicted by all his friends and neighbours when he told them that the 'people from the TV' are coming.

The first rule is to use the minimum of everything. The minimum number of people, minimum equipment, minimum lighting, and cause the minimum disturbance. Leave people and equipment not essential to the actual shooting outside in the car and forbid them to come near the shooting area.

Camouflage yourself and equipment to look like the people you have to film. If they all wear ties, all of the crew, including the driver and the electricians if they are to mix, should also wear ties, etc. Don't wear jeans to interview the Pope, the Queen or the President of the United States.

Don't use extra lighting unless absolutely necessary. It is much better to use fast lenses; a *f*1.3 lens is almost two stops faster than one whose maximum aperture is *f*2.5.

Be inconspicuous with microphones. The thought that their words are being recorded is very daunting to most people. The best microphones are those which can be used from such a distance that they do not give a fenced-in feeling.

Be polite

Take it easy, especially in the beginning

Display the minimum amount of equipment

Don't disturb the subject's environment

Use fast lenses rather than supplementary lighting

Merge into the style of dress and behaviour of the subject

Be inconspicuous with microphones

Refrain from using incomprehensible technical jargon

Minimise the number of people watching the subject perform

Relax .......... the crew and the subject

# Tricks of the Newsreel Trade

Each branch of film-making has its special skills, not least the newsreel cameraman.

## Typical ploys

Have confidence. If an official looks at you as if you have no right to be there, look straight back at him as though you have.

Dress as though you belong to the scene. In a discotheque wear jeans and a flowered shirt, in an office a jacket and tie. In someone's home respect it as they do. Respect for other people's scene pays big dividends in terms of co-operation and facilities.

Think ahead. Work out what is going to happen next, where and when and be there. Don't worry about what the competition is doing, make them worry about you.

When filming an event where certain competitors are more important than others, write their numbers on the back of your hand. It is quicker than looking them up in a programme.

Communication is your business. Keep in touch. Listen to news bulletins, read newspapers, phone the office, talk to people, write full details about what you have shot.

Always have spare film, battery and cable near at hand. Anything may happen at any time and nothing is worse than being unable to shoot.
Fast lenses take pictures when it is too dark to film by any other means.

Leave about 20 or 30 sec of film in a magazine at the end of a roll as a reserve supply to fall back upon in an emergency.

When someone or something is going to approach the camera decide beforehand if and where you will shoot in close-up, on the long end of a zoom, and pre-focus on that spot. The remainder may be shot in wide angle, letting the depth of field take care of the focus.

Use a stop watch when filming motor racing. If you know the lap times, you can switch on just before a certain competitor arrives, can work out who is gaining on whom and calculate how many laps have been completed since the start of the race.

A tripod is the perfect aid to steady pictures, but to rest one's elbows on a ledge, against a building or even on a knee when kneeling is better than nothing.

Approximate focus and aperture settings which can be seen by the spare eye while shooting can be very reassuring.

Identify the beginning of every roll, and put synchronism marks on every sync scene to keep the editor your friend.

When working in a dangerous situation and joined to the sound recordist by a pulse cable, put a pull away plug in the line so that you can both jump your own ways.

Remember, in the history of newsreel cinematography, the shot that was worth dying for, has yet to be taken.

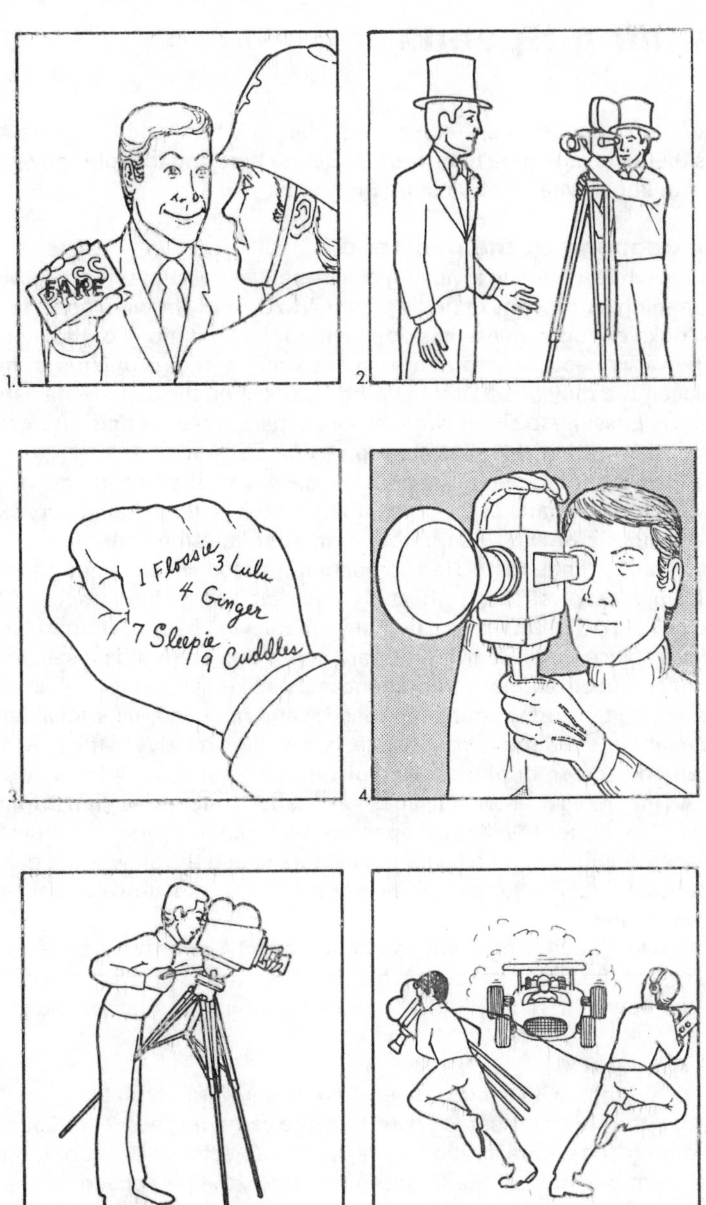


TRICKS OF THE NEWSREEL TRADE

1. Have confidence; 2. Try to become a 'part of the scene'; 3. List the favourites for quick reference; 4. Use ultra-wide aperture lenses in low light; 5. Use a tripod as an aid to steady pictures; 6. Don't become 'tied' to your work.

# Panning Speeds

Cinematography is an art and, as such, has few hard and fast rules. What rules there are, are there to be broken. Before breaking the rules however, it helps to know what the rules are in the first place.

### Rule applying to the pan handle

Panning while following a moving object is always acceptable on the screen but panning as a means of getting from A to B, or of showing both A and B, may look very ugly unless it is done within the limitations of the medium.

When it is necessary to pan across a static scene, problems of image displacement may arise. Before doing such a shot, the cameraman should ask if it is possible to shoot two, or more separate scenes and cut from one to the other, rather than pan through a static scene.

If a pan is preferable, perhaps because it is desired to maintain a link between the two parts of the scene, then it must be done either very slowly or very quickly (a whip pan) rather than at an intermediate pace.

There are definite restrictions on panning speed which, if ignored result in an unpleasant strobing effect. Tables detailing panning speeds have been calculated on the premise that human eyes, being set apart, see an object displaced by just over 7° of arc and if this amount is exceeded the brain is confused, cannot comprehend and dislikes what it sees. Acceptability is dependent upon: panning speed, camera speed, lens focal length, shutter angle, the direction the shutter is travelling relative to the pan (or tilt) movement, screen brightness and picture contrast ratio. A low key scene without harsh black and white lines withstands faster pans than brightly lit scenes. The larger the shutter opening the less the chance of unpleasant displacement effects. The faster the camera speed the slower the panning speed must be. The longer the lens focal length the slower must be the panning speed.

In practice such tables are too complicated to consult as they really require that the length of the pan be measured with a protractor and timed with a stopwatch. In the field, this would hold up the unit for too long.

### Will it strobe?

Rule of thumb: with static subjects (with any film format or lens focal length) if the shutter is set at about 180°, the camera speed 24 or 25fps and the lighting and contrast ratios average, if a subject takes 5 sec or longer to travel from one side of the finder to the other, the pan speed is safe.

POSSIBLE PANNING SPEEDS

1. Following a moving object is always allowable; 2. Fast whip pan from one part of a scene to another is allowable as is a slow pan where any one point of the scene takes at least five seconds to pass from one edge of the picture to another; 3. Wide angle shots (A) can be panned faster than long lens set-ups (B).

# Zooming and Dollying

Zooming does not give the same visual effect as dollying.

### What a zoom lens does

Imagine a person 10ft from the camera; behind him, 10ft (3m) further away, some detail, unimportant to the shot, and beyond, stretching away to infinity, the background.

With the lens set on wide angle, the foreground man is full length, mid-range objects half that size and the background has depth. The scene is roughly, as perceived by the eye. If we adjust to a focal length ten times as long, the person in the foreground grows until seen in close-up—with just a head and shoulders filling the screen. Behind him, in the middle distance, the unimportant detail is magnified until it is almost as important as the foreground subject and the background comes crashing inwards until only a small central portion fills the screen, having no natural depth or perspective. Furthermore, when the zoom lens was set on wide angle everything was acceptably in focus from foreground to background. When zoomed in, the foreground is sharp, the middle distance is so-so and the background, such as it is, is out of focus. There are times when such an effect is acceptable, even desirable.

### What dollying does

Imagine the same shot as before, but instead of zooming-in, the camera is dollied-in until the person is 2–3 feet from the camera. The person is large in picture, as before, the unimportant detail in the middle distance is slightly larger but less important relative to the foreground than before and the background remains about the same. The shot retains a natural perspective and although some focus may be lost on the background, it still remains recognisable.

### The compromise

Zooming is an unnatural effect. Sometimes it is unavoidable, and as such may sometimes be used to advantage, but on those occasions, when the director has a choice and does not want to draw attention to the technology, it is very often preferable to dolly than to zoom.

Zoom effects may be hidden, or minimised, by combining them either with a dolly or a pan or by doing all three movements simultaneously.

On news and documentary coverage it is sometimes possible to do a crash zoom, often while there is a pause in the sound, and persuade the editor to take out the spoiled frames at the cutting stage. In this way, the transition may be made to look like a straight cut.

ZOOMING AND DOLLYING

The effect of zooming (1, 2, 3, 4). The effect of dollying (5, 6, 7, 8). Note how the background proportions are totally different.

*Cameramen have made their reputation by hand-holding and kept it by using a tripod.*

# Hand-Holding a Camera Statically

Great importance is placed upon skill in hand holding a camera while shooting for the so-called 'mobile camera' mode of film-making.

### Hand-held vs. fixed camera support

The camera may be hand-held in emulation of a tripod and then moved to another position without need to set up the reverse angle as a separate shot. Sometimes the hand-held camera can be taken into positions inaccessible or inconvenient for tripod or dolly operation. Occasionally a camera is hand-held instead of being tripod-mounted for reasons of pure laziness and unprofessionalism. Whenever it is possible to support a camera mechanically, this form of operation is always to be preferred as any shaking of the camera may draw the viewers attention to the technique rather than the subject of the picture.

### Maximum steadiness

A camera may be held most steadily providing as rigid support as possible, and restricting image and movement magnification if it is supported on the cameraman's shoulder and held firmly against the side of his head by hands which, themselves, are supported and steadied by elbows pressed hard into his waist. Cameras of poor ergonomic design, if they must be hand-held, are best supported by a shoulder bracket.

### Choice of lenses

To achieve the steadiest looking picture on the screen wide angle lenses should be used, where possible, in preference to those of longer focal length. It is very easy for a cameraman to be tempted, particularly with zoom lenses, into being over-ambitious in the focal length of lens he uses. He should always look carefully at his material on the screen and reach a conclusion as to what focal length he should confine himself to for the standard he seeks to achieve—and exceed it only in a dire emergency, knowing how poor the result will be.

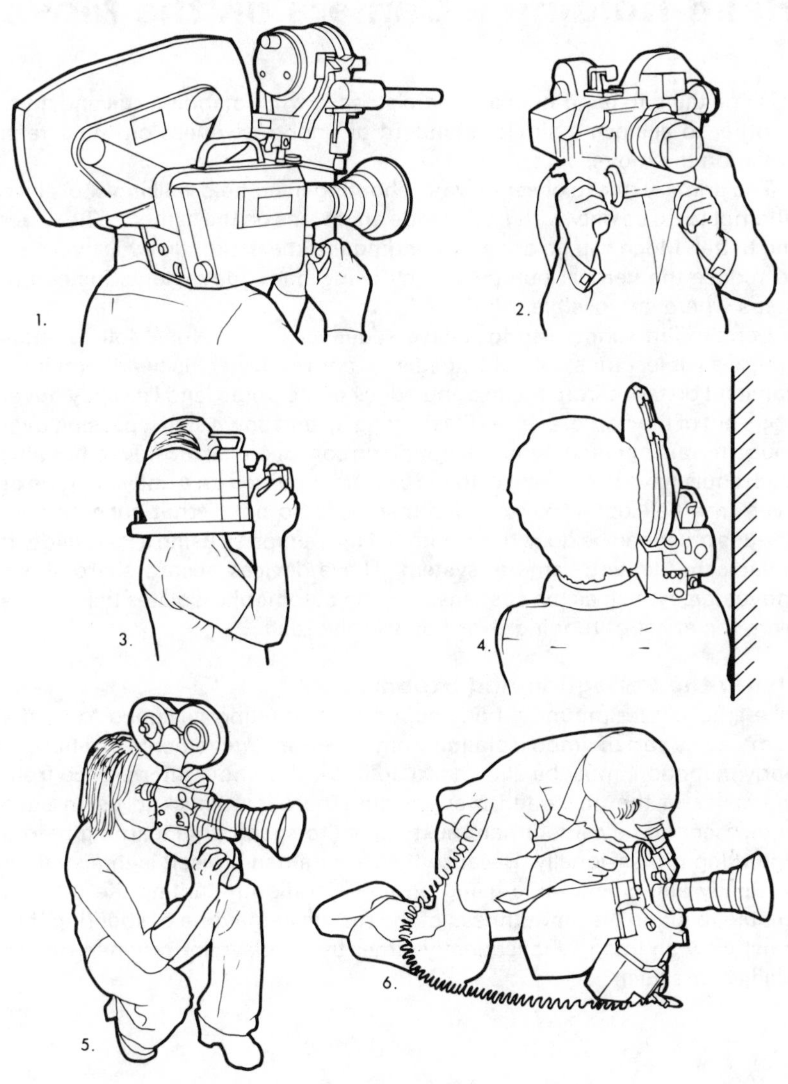


STEADY HAND-HOLDING STATICALLY

For steady hand-holding, cameras should sit firmly on the operator's shoulder (1). He should grip the camera firmly (2), and dig his elbows well into his sides (3). For additional support he may lean against a wall (4), kneel (5), or rest the camera on the ground (6).

*Impersonating a dolly.*

# Hand-Holding a Camera on the Move

It is one thing to hand-hold a camera steadily while standing still and quite another to achieve a similar standard of unobtrusiveness on the screen while on the move.

The primary rule to observe when shooting hand-held walking shots is to attempt to do so only with a very wide angle lens on the camera. The wider the better. Much can be done by bending the knees slightly while walking, to reduce the vertical bumps and by using body movements instead of steps wherever possible.

Certain film scripts call for chase sequences and involve follow shots where a subject must be held steadily in frame. Ideally his head remains a constant distance from the top and edges of the frame, and certainly never pops out of the frame, even while running up and down stairs, passing over rough terrain or being bumped around on some conveyance. Traditionally, such shots have been done with the camera mounted on a dolly or crane or even in a helicopter. Now if circumstances do not permit such camera set-ups much can be done to smooth out the bumps by using a Panaglide or Steadicam 'floating camera' system. These devices absorb sharp shock movements much as the suspension of an automobile isolates the chassis from the effect of running over holes in the road.

### Using the Panaglide and Steadicam

Like a helicopter mount, a body-supported mounting designed to hold a camera in a sort of limbo isolation from the erratic movements of a human body in motion, must be allowed to do its work without interference from the cause of the unwanted movements. There is a knack to using such equipment and a cameraman must expect to spend time learning before operating professionally. Because the cameraman himself is in a state of unsteady animation while using an anti-vibration mounting, he may be unable to judge the smoothness of the results at the time of shooting, but must view an image subsequently either by TV playback or when rushes (dailies) are seen.

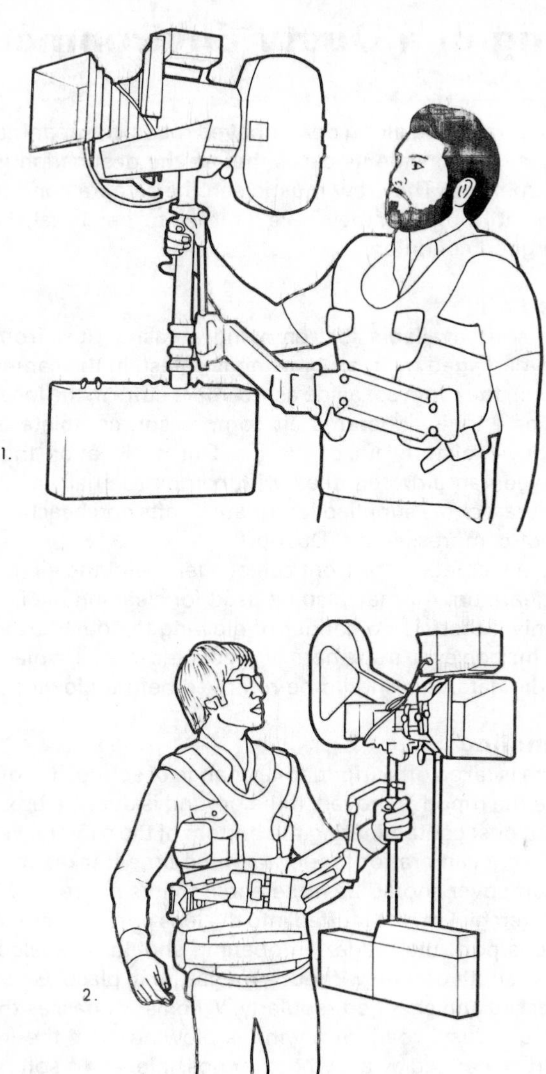

STEADY HAND-HOLDING ON THE MOVE

1. Panavision Panaglide; 2. Cinema Products Steadicam.

*A problem encountered in vast areas of the world.*

# Shooting in a Dusty Environment

On location and travelling along dry, unpaved roads, the gear may become covered in dust, even inside its cases, before the destination is reached. Fine dust is a menace. The crew must clean the camera continually from early morning until long after they have returned to their hotel, if they are to maintain it in good condition.

## Compressed air

Of all the methods available for removing abrasive dust from surfaces which may be damaged by it, air blowing is safest. In the camera maintenance shop, or in the room set aside as a camera store in the location hotel, there should be a mains-operated air compressor, complete with a unit which takes any moisture out of the air. Out on location the choice is probably between an airbottle filled with compressed air, a 12v battery-operated compressor (as supplied for inflating rafts and beach mattresses) and aerosols of compressed air (Dust-off).

When using an air jet on the front cell of a lens, aim the air jet sideways rather than square-on. Air may also be used for cleaning the inside of the camera but only if there is no danger of blowing the dust further into the body where it may do even more harm. If a gelatine filter is in place between the lens and the gate, this should be removed before blowing out.

## General handling

When a camera is taken off a tripod for a change of set-up, it is often put on the floor while the tripod is moved. If the ground is dusty, a box should be used to prevent dust contaminating the bottom of the camera. When a lens is removed from a camera for longer than the time it takes to fit another lens, a lens port cover should be fitted and if that is not readily to hand, a duster or handkerchief can be stuffed into the lens port as a temporary seal. As with the lens port, the magazine opening should be sealed if it is to remain for any length of time without a magazine in place. Gelatine filters should be checked and changed regularly. When a car passes the camera, travelling along a dusty road, or a wind is blowing dust, the front of the camera should be panned away, wherever possible, and a soft brush used to remove dust from the exterior. No surplus oil should be allowed to remain on the surface of any equipment as, inevitably, any dust will adhere to it.

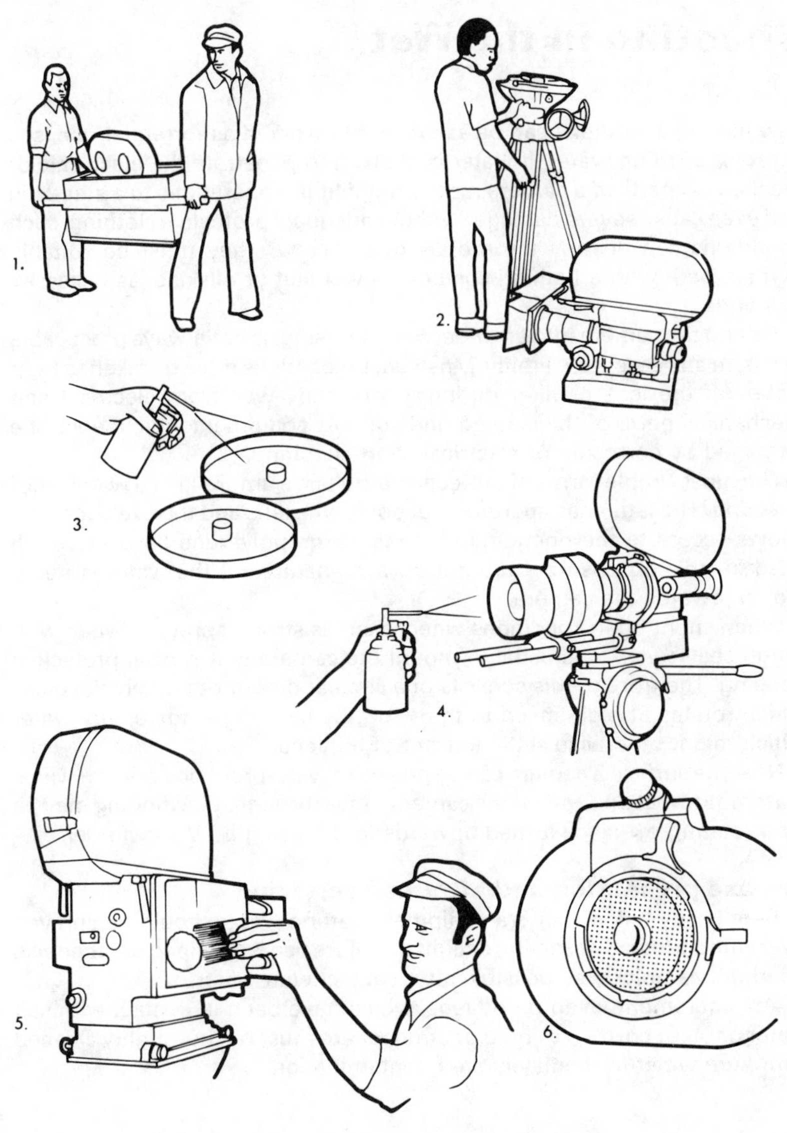


SHOOTING IN A DUSTY ENVIRONMENT

1. Protect camera when carrying it around; 2. Put camera on a box when moving tripod; 3. Use compressed air to clean magazines; 4. Use 'dust-off' to clean lenses on location; 5. Use brush to remove dust from camera body; 6. Never leave lens port uncovered.

*An environmental hazard with great powers of penetration.*

# Shooting in the Wet

Few filming conditions can be as difficult to work in as torrential rain, sea spray or swirling water. If water is allowed to penetrate the electronic or mechanical parts of a camera it can bring filming operations to a standstill and even cause severe damage. Without adequate protective clothing, such conditions can destroy the morale of the crew. They must be suitably garbed, be it with a bathing costume, a wet suit or oilskins, as necessity demands.

Putting the camera into an underwater housing, is not always practicable and is, in any case, very limiting. Instead, precautions must be taken to keep water off the lens or filter during a take and away from electrical and mechanical parts of the camera and support equipment which would be damaged by corrosion or electrical short-circuiting.

The most simple forms of protection are a large umbrella or a waterproof covering. The latter has apertures cut out for the lens and the eyepiece, and allows access to the operating controls. Large polythene bags are often used for this purpose but are not always adequate if the water is being blown towards the camera.

When shooting in conditions where there is strong spray, as when with rough sea scenes, it is usual to mount the camera in a special protective housing. The front of this consists of a circular disc or optionally flat glass which rotates at high speed to throw off, by centrifugal force, any water which might otherwise settle in front of the lens.

Near the surface a camera can be put into a waterproof box bottom with a glass window at one end for the camera to see through. Viewfinding may be by a rotating eyepiece turned upwards, or by using a TV viewfinder.

## Pre-use precaution and after-use servicing

If there is any risk of camera equipment getting wet it should be sprayed over with a protective coating or light oil before use, and wiped or otherwise dried out as quickly as possible after contact with water.

Any equipment, even tripod legs, which have been in contact with salt water or other corrosive liquid or atmosphere must be thoroughly cleaned with pure water immediately after contamination.

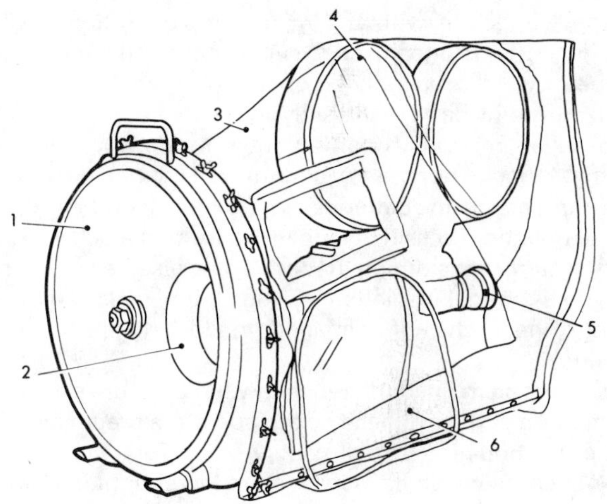


SPINNING DISC RAIN DEFLECTOR

1. Disc of glass, mounted on a central spigot, which rotates at high speed. Any water falling on the surface is thrown off by centrifugal force; 2. Camera lens; 3. Transparent waterproof cover; 4. Camera; 5. Camera eyepiece protrudes through an opening in the cover; 6. Flap to permit access to focus and zoom controls.

# Shooting in the Cold

Extremes of cold may be encountered when working in the Arctic, up high mountains or when travelling in high-flying aircraft. Down to 0°F (−15°C) cameras and filmstock should require little special preparation.

## General precautions

Cameras are often fitted with internal heaters or may be supplied with electrically heated barneys which should be switched on prior to starting, leaving sufficient time for the camera to reach a working temperature. The camera may have to be run, without film, until it is operating satisfactorily.

A portable generator is often used to supply power to internal heaters or barneys, some of which may only be operated off an AC supply.

Below freezing point batteries rapidly loose their capacity to produce power and because bearings become tighter, more power is required to operate a camera than at normal temperatures. It may be necessary to use batteries of a higher capacity. Batteries may be kept at an effective temperature by the use of battery cells made up in the form of a belt which may be worn beneath waterproof clothing. Batteries may also be contained within a box lined with a thermal insulation. As the battery is used it dissipates a slight amount of heat which if contained, may be reapplied to conserve battery efficiency.

An infuriating problem encountered in low temperature operation is the constant misting up of the eyepiece by the operators breath and the warm temperature of his body in close proximity.

Cameras and lenses which have been stored overnight in a warm room should be taken into the cold well before they are required and slowly allowed to reach the operating temperature. Preferably, cameras which must operate in the cold should *not* be kept in a warm room overnight.

When a cold camera is brought into a warm room for immediate operation, the lenses invariably mist up internally. Exposing them to the warmth of a lamp accelerates the process of normalisation.

When a camera is brought in out of the cold and allowed to normalise slowly, it should be kept in a plastic bag to provide alternative surfaces for the condensed moisture to precipate upon. This will reduce corrosion and remove water which would freeze when next the camera is exposed to the cold.

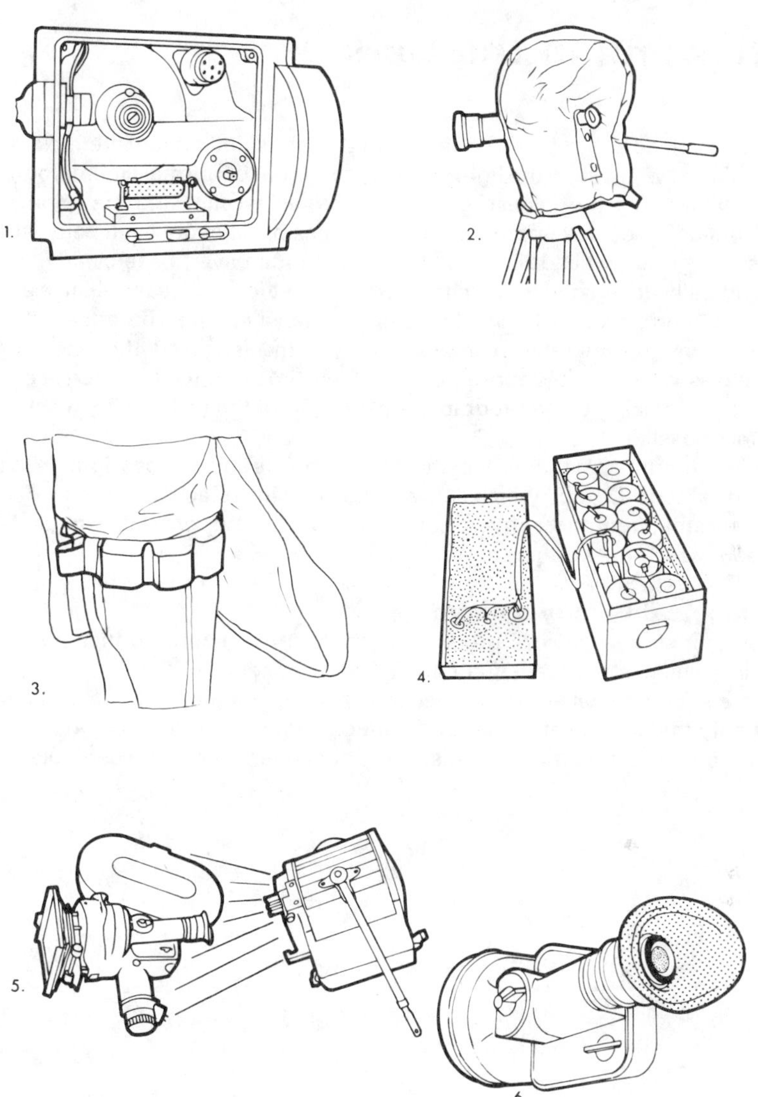


SHOOTING IN THE COLD

1. Heater element placed inside a Mitchell BNCR camera below the main drive cross shaft; 2. Barney with internal heater elements placed over a camera; 3. Battery belt worn under clothing; 4. Battery case lined with heat insulating material; 5. Lamp used to warm camera eyepiece; 6. Camera eyepiece may have anti-mist coating or have a small heater element attached.

# Shooting in the Dark

It is possible to shoot film with the aid of a 'night-sight' image intensifier or amplifier at well below the light intensity required for normal cinematography, even using the fastest lenses and the fastest film and the greatest amount of forced development. It is in fact feasible to shoot even below the level of light in which it is possible for the human eye to see clearly.

Originally developed as a military requirement for night surveillance and gun sighting, a small lightweight image intensifier may be attached to almost any 16mm reflex camera, between the lens and the film, and increases exposure by approximately 1000 times, a 10-stop increase in exposure, making cinematography by the light of the (celestial type) stars almost possible.

A lens mounted at the front of the device collects and focuses light on to a photo cathode TV tube which converts the light into electrical energy. This is then amplified and converted back to light in the form of a small TV screen, which may be photographed.

## Image quality may be poor

The most simple devices produce a greenish monochrome picture of a quality which is quite adequate for TV news usage.

Care should be taken when selecting an image intensifier to ensure that not only the amplification is good enough for the requirements, but also that the image definition and distortion characteristics are acceptable.

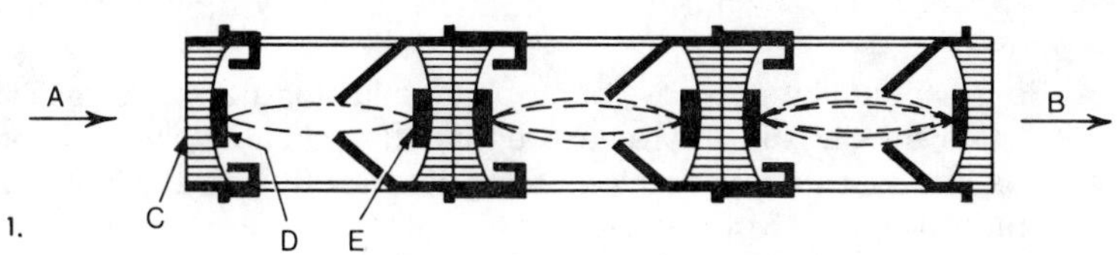


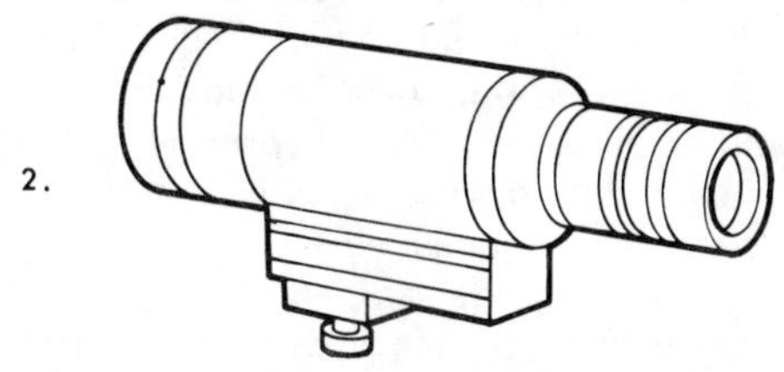


SHOOTING IN THE DARK

1. Three-stage image intensifier: A. Image in; B. Image out; C. Fibre Optics Plate; D. Photocathode; E. Luminescent screen. 2. External view.

# Shooting for video

When filming specifically for video it is as well to recognise the limitations of this form of presentation and tailor the images to suit.

## TV cut-off

The areas scanned by a telecine machine and subsequently displayed by the average domestic TV receiver are significantly smaller than the frame areas of the original film and to allow for this, film cameras used for filming video material should be fitted with a ground glass which has 'TV safe action area' markings. This area is about 13 per cent less in width and 11 per cent less in height than the 35mm Academy or 16mm 'projector' frame.

The area actually seen by many badly adjusted domestic receivers is very often smaller still and to ensure that titles remain fully comprehensible an extra margin must be followed. The 'safe title area' is about 23 per cent narrower and 15 per cent less in height than the 'projector' frames. TV safe action and TV safe title area ground glass sizes are discussed in *Camera and Lighting Equipment, Choice and Technique* page 28.

## Visual limitations

Of the 625 lines transmitted on a 50Hz TV system only about 560 are seen on a domestic TV set and on a 525 line/60Hz system the number of active lines is only about 470, therefore only this number of bits of information may be resolved. The horizontal information is similarly limited. By comparison, an optically projected image can define considerably smaller detail and sharper lines. For this reason wide angle scenes, containing a mass of detail and small patterns and incorporating fine lines, reproduce poorly on video as compared with the cinema and other projections and should be avoided whenever possible.

Another reason why the video medium is more ideally suited to close-ups is because viewers tend to sit further away from TV screens than from cinema screens. On average, viewers at home sit about 5 or 6 screen widths away from a domestic TV set, whereas if they were to sit that far away from an anamorphic cinema screen, very often they would need to be out in the street.

SHOOTING FOR VIDEO

*Good for video:* 1. Clothes with colours which differ in tone as well as in hue; 3. Big close-ups; 5. People against mid-tone background; 7. People in centre of screen; 9. Titles should be in centre of screen.

*Bad for video:* 2. Small check patterns on suits and costumes; 4. Masses of small detail; 6. People against light backgrounds; 8. People on edges of screen; 10. Titles should not be at the screen edges.

# Shooting Colour Film for Colour TV

Often, a mainly electronically generated video production requires insert shots on film. This almost invariably allows greater creative scope.

Film may be preferred to electronic cameras for exterior scenes and sequences shot at distant locations, where a small film crew is more economical. These scenes may be cut into any drama, comedy, documentary or other types of programme otherwise shot in a studio.

Ideally, the transition from one medium to another should be done without any noticeable change in quality.

## Matching two systems

Much of the smoothness of the intermatching depends upon the facilities and expertise of the telecine department of the video company. But, initially, the control is with the programme producer and the film cameraman.

Film is primarily intended for subtractive colour projection. Inevitably the hues, contrast, saturation, and luminance of the colours as well as the definition, grain and image steadiness characteristics of film and electronic cameras differ.

## Minimising differences

When matching film to video, and desiring to achieve an unnoticeable transition from one medium to the other, the following colours of scenery, props and clothing should be avoided, preferably from the planning and design stages.

1. Deep reds (film reproduces them better than video). 2. Saturated cyans (video copes better with these blue-green colours). 3. Any brightly-lit saturated colours (video is less adequate than film). 4. Saturated colours in the shadows (film is inferior to video here). 5. In particular, deep purples should be avoided as these respond quite differently to both film and video cameras.

In general, the best results are achieved with colours and lighting of mid-saturation and contrast. Avoid contre-jour (against-the-light) filming where colour saturation matching is important. Definition, grain and image steadiness differences could be minimised by using 35mm film. But with the improved filmstocks and lenses introduced in the mid 70's, 16mm film shot to feature film production standards is totally acceptable even for drama inserts and international prestige programmes. Film intended for video transfer may be printed on a special low contrast print stock if preferred.

The film cameraman, the film laboratory, the broadcast company's Head of Film and the Telecine Supervisor should all liaise (even if only on the telephone) so that each understands what the other is trying to achieve. This is particularly important with scenes incorporating mood lighting and special effects.

FILM AND VIDEO FOR COLOUR TV

*Incompatible.* (Film more faithful): 1. Deep red colours; 3. Brightly lit saturated colours; 5. High key backgrounds. (Video more faithful): 2. Cyan colours; 4. Saturated colours in the shade. (Neither better): 6. Contre-jour lighting. 7. Deep purple colours.

*Compatible:* 8. Mid-scale colour saturation and contrast.

# Shooting with Metal Halide Lighting

Metal halide lighting is an efficient form of illumination giving off three to four times as many lumens of light per watt of electricity consumed compared with a normal bulb with a tungsten filament.

MH lighting must be operated off an AC supply (never DC) and will pulse at twice the supply frequency. A ballast unit must be in circuit between the mains supply and the lamp-head.

To ensure flicker-free filming the camera must either be operated at a speed which is divisible into the lighting frequency (25 fps with 50 Hz supplies) or in conditions where the camera speed and shutter opening (exposure time) and lighting supply frequency are compatible.

Alternatively, MH lights may be used with 'square wave' or 'high frequency' units specifically designed to give flicker-free filming from any supply at any camera speed and with any shutter opening.

## Settings for flicker-free MH operation

Here is the way to calculate compatible settings.

**Two light pulses per exposure:**

Shutter angle = camera speed × 360 ÷ supply frequency
Supply frequency = camera speed × 360 ÷ shutter angle
Camera speed = supply frequency × shutter angle ÷ 360

**Any number of light pulses per exposure:**

**50 Hz supply**

Shutter angle = camera speed × 3.6 × number of light pulses per exposure
Camera speed = shutter angle ÷ 3.6 ÷ number of light pulses per exposure

**60 Hz supply**

Shutter angle = camera speed × 3.0 × number of light pulses per exposure
Camera speed = shutter angle ÷ 3.0 ÷ number of light pulses per exposure

*Relationship of 24 fps camera to 60 Hz supply.* If a camera running at precisely 24 fps photographs a subject illuminated by MH lamps operating off a 60 Hz supply, it will be illuminated by 2½ light pulses per exposure.

Assuming the camera speed to be precisely 24 fps, any variation in the power supply frequency will result in a variation in the picture luminance similar to that produced by an exposure change of 0.4 stops. The time taken for one cycle of variation will depend upon the extent of the variation viz:

| Power supply | Fluctuation cycle |
|---|---|
| 60 Hz ± 0.0025 | 200 seconds |
| 60 Hz ± 0.005 | 100 seconds |
| 60 Hz ± 0.01 | 50 seconds |
| 60 Hz ± 0.02 | 25 seconds |
| 60 Hz ± 0.05 | 10 seconds |

A variation in camera frequency will also cause a 0.4 stop variation in exposure. Assuming the 60 Hz supply to be precise, the cycle of exposure variation caused by a camera running at 24 ± 0.000003 would be approximately 4½ minutes. However, if the errors were combined, a 60 Hz ± 0.0025 power supply light for a camera running at 24 ± 0.000003 could result in an exposure variation every 116 seconds.

## FLICKER-FREE LIGHTING 'WINDOWS' FOR METAL HALIDE LIGHTING

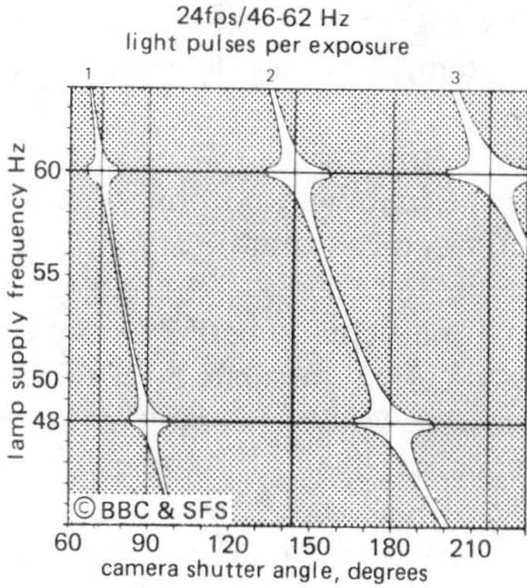


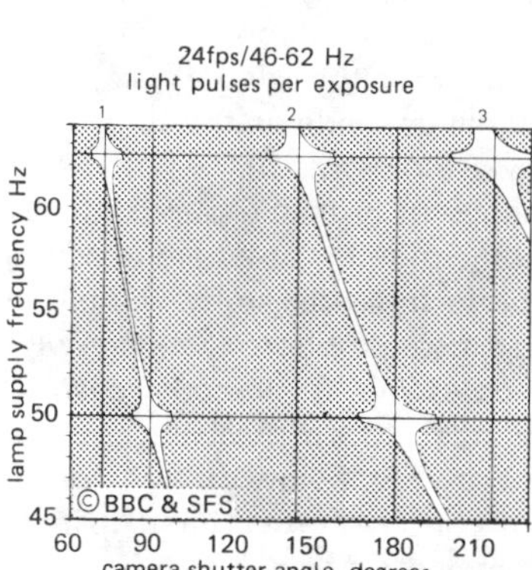


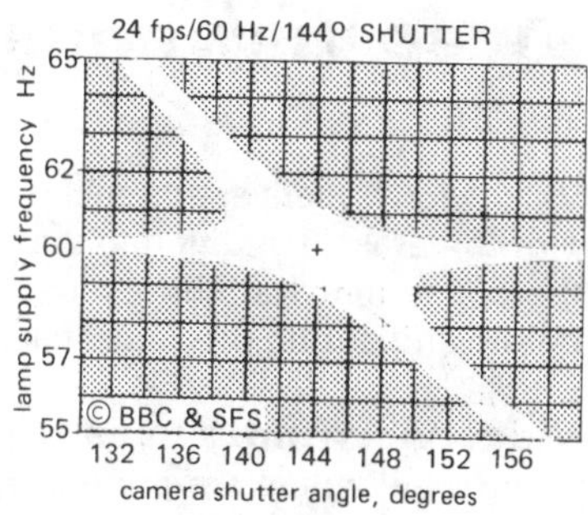


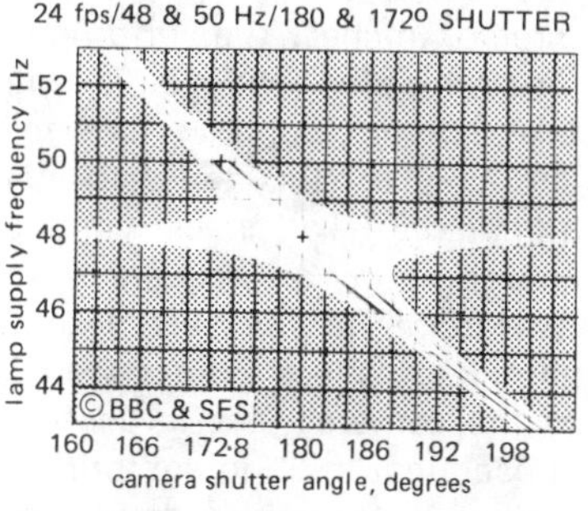


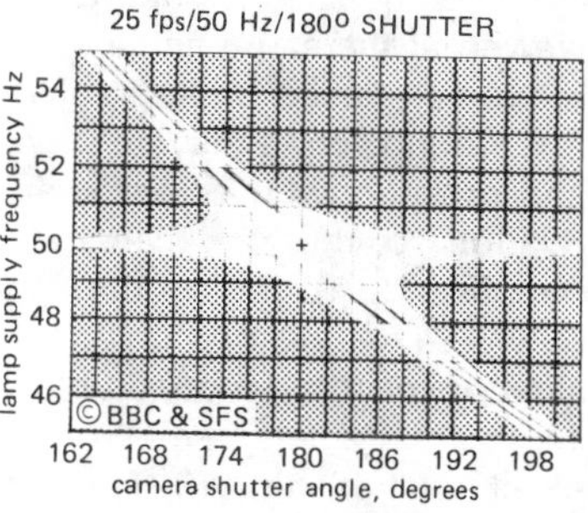


### OPTIMUM SUPPLY FREQUENCIES AT 24 FPS FOR VARIOUS SHUTTER OPENINGS

| Shutter Angle | | Optimum Supply |
|---|---|---|
| 144° | Optimum setting for 24 fps/60 Hz operation | 60 Hz |
| 165° | | 52.4 Hz |
| 170° | | 50.8 Hz |
| 172.8° | Optimum setting for 24 fps/50 Hz operation | 50 Hz |
| 175° | | 49.4 Hz |
| 180° | Optimum setting for 25 fps/50 Hz operation | 48 Hz |
| 200° | | 43.2 Hz |

At 24 fps/48 Hz or 25 fps/50 Hz EXACTLY, any shutter angle is possible. At 24 fps/60 Hz/144° shutter, 24 fps/ 48 Hz/180° and 25 fps/50 Hz/180° shutter the tolerances are comparatively wide compared with other combinations.

Tolerances for CSI and CID lights are considerably greater than for HMI.

# Video Viewfinders

Video viewfinders are often incorporated into the reflex viewing system of film cameras, in addition to the optical viewfinder to enable others to view the scene and framing exactly as seen by the camera operator, either before and during the actual take, or subsequently from a video recording.

## Video viewfinder uses

Video viewfinders are principally used for the following purposes:
1. To enable the director to view a take precisely, so that he can re-view doubtful takes and very often save shooting re-takes. 2. To enable the director to check the performance of artists—particularly useful when the director is also acting. 3. To reduce processing and print costs by checking takes before sending film to the laboratory. 4. When multi-camera shooting, to stop and start cameras to save expense. 5. As a guide to editing. 6. To check continuity. 7. To check lip movement and choreography on sequences shot to play-back. 8. To enable specialists (wardrobe, hairdresser, make-up, props) to view a scene as seen by the camera, to check their concerns. 9. To place objects accurately in relation to the camera. This is useful when shooting pack (product) shots for subsequent multi-image printing. If desirable, each component of a shot may have its position drawn on the face of the monitor using a wax pencil (when doing so, it is important that any markings made on the screen are viewed at the same angle at which they were drawn). 10. For travelling matte composite photography. Video images of both the plate and the foreground action may be superimposed on a monitor as an aid to acting and operating. 11. To assist an artist to perform a precise movement in relation to the film camera, particularly useful when working with puppets. 12. To reduce the number of people who, for various reasons, may wish to look through the camera viewfinder. 13. To check the effect of slow or speed-up motion by playing back the VTR at slow or high speed. 14. To show a held frame from one take while another take is lined up in an exact relationship. 15. For checking all manner of special effects for credibility. 16. To check technical aspects of a take—lighting, operation, focus. etc. 17. To enable a camera to be placed in a remote, difficult, or dangerous position and be operated by remote control. 18. Post sync dialogue recording.

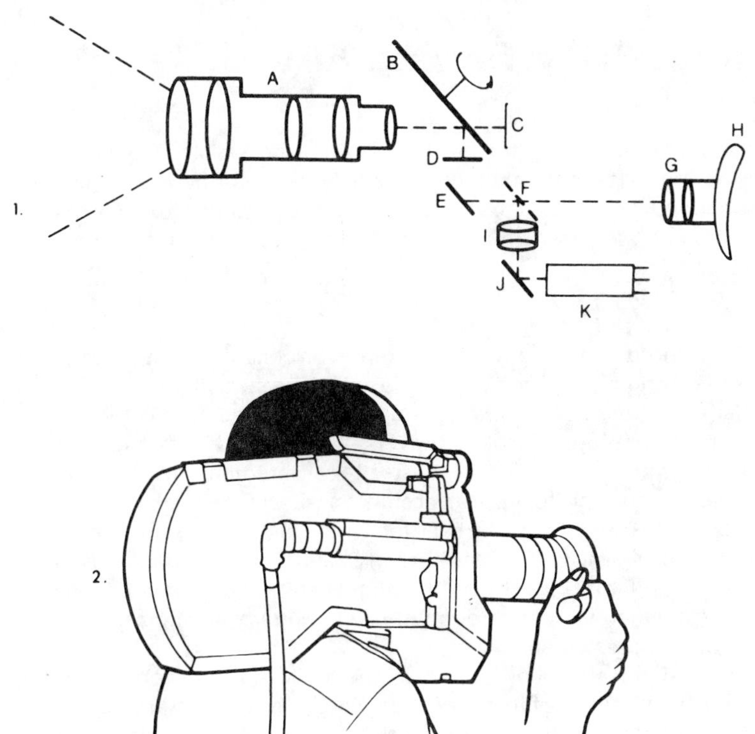


VIDEO VIEWFINDER

1. Typical video viewfinder layout: A. Zoom lens; B. Mirror shutter; C. Film plane; D. Ground glass; E. Mirror; F. Partial reflecting mirror; G. Optical Viewfinder; H. Operator's eyepiece; I. Video camera optics; J. Mirror; K. Video camera.
2. Aaton 16mm camera, with built-in video viewfinder, in use.

Under normal circumstances the video camera 'sees' the scene via the spinning reflex mirror (B) during the film pull-down period and shares the available light with the optical viewfinder (G) through a partial reflecting mirror (F). On some cameras which have a focal-plane shutter in addition to the rotating mirror (B), it is possible to replace this mirror with a partial reflecting mirror to give a higher-quality flicker-free image.

Where the video camera is being used on a remotely controlled film camera and the camera operator has no need to look through the optical viewfinder, the partial reflecting mirror (F) can be replaced by a 100 per cent reflecting mirror to increase the amount of light to the video camera.

# Using a Viewfinding Filter

The contrast range which the motion picture process is capable of reproducing is much less than that which is acceptable to the human eye. An allowance must be made for the fact that the eye can see details in both highlights and shadows which cannot be reproduced on either a motion picture or TV screen.

## Colour contrast viewing filter

When a cameraman is lighting an interior or considering an interior he knows that certain areas do not photograph as he sees them. For instance, an artist's eye sockets which appear dark because the light is too directly overhead, will photograph black. Conversely, hot reflections dominate their surroundings and cause flare. All these, the cameraman can control. But to do that he must be able to make an accurate assessment. Looking through a colour contrast viewing filter he sees the whole picture darkened down and reduced in contrast. Dark areas go very dark and lack detail, as when reproduced on film. Uncontrolled highlights stand out and if they appear 'burned out' when viewed through a viewing filter they are certain to be burned out on the screen.

Viewing filters are incorporated in the viewfinder systems of many cameras used for feature film production. Particularly handy are those fitted in cameras which also incorporate a viewfinder magnifying system used to enlarge a small area of the scene for individual examination.

## As an aid to setting lighting

Viewing glasses are used by cameramen and gaffers when setting lighting. When a lamp is looked at from the position of the subject, particularly if it is fitted with a lens, it is possible to tell exactly where the beam is centred and the degree of flood or spot effect coming from it. Unless a viewing glass is used for this purpose the cameraman or gaffer temporarily blinds himself and can not judge any other lighting effect again until his eyes have normalised, which may take some time.

A viewing filter is also used when looking up into the sky to assess when the sun is going to shine and for how long.

## Pan glass

In the days before colour film, cameramen used viewing filters to help judge the colour separation, or otherwise, likely to be reproduced when a multicoloured subject was photographed with a monochrome emulsion. There was one type for orthochromatic film and another for panchromatic. The latter was often called a 'pan glass'. If a B & W film is used today it is still important to use the correct viewing glass. Those intended for colour film are not suitable for monochrome.

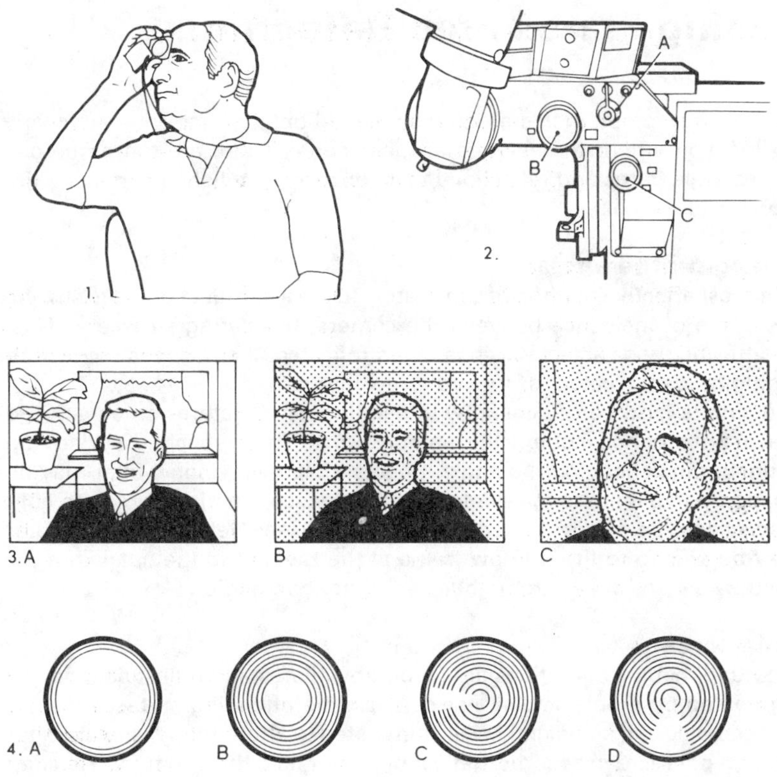


USING A VIEWFINDER FILTER

1. Cameraman looking through viewing filter; 2. Viewing controls of a Panaflex camera: A. Contrast filters; B. Image magnification; C. Image normal, de-anamorphosed or eyepiece closed. 3. Scene as seen through camera viewfinder: A. Normal; B. With contrast filter; C. With contrast filter and image magnification. 4. Fresnel lamp as seen through viewfinding filter: A. Spot; B. Full flood; C. Turned to one side; D. Tilted upwards.

# Losing Flares and Reflections

Among the cameraman's problems when shooting interiors, particularly on location where the settings are less controllable, are the numerous undesirable flares and reflections that come from polished surfaces, windows, pictures, etc.

## Changing the angle

The most effective method of eliminating flares and reflections is to change the angle of incidence between the camera, the flaring surface and the lamp, window or object which is being reflected. If any one is moved the highlight is eradicated, or moved elsewhere.

A simple method of dealing with paintings and pictures on a wall is to place a matchbox behind the painting to tip the reflecting surface up slightly, thus changing the angle. Sheets of acrylic window filter may be tipped up to eliminate the single reflection which remains from each light when this type of daylight colour control is used. Gelatine window filter may be pinned to the window frame at the top and to the outside of the window sill and the bottom, thus setting it at an angle.

## Pola screens

Provided that the conditions are favourable, flares and reflections may be eliminated by a pola screen. For success, the offending surface must be non-metallic and sufficiently well illuminated to afford the 1.5 stops of light which a pola screen absorbs. Reflections at, or near the optimum reflecting angle of 37° are most effectively counted.

Reflections from metal surfaces and from any other surface where the critical angle of 37° cannot be achieved may be eliminated by polarising the light source as well as the camera lens. Pola screens fitted to the lamps must be set at right angles to the pola screen on the camera lens. This method is particularly useful for filming objects with curved or angular surfaces, products wrapped in transparent packaging and subjects set behind glass.

## Anti-flare

A dulling spray, applied from an aerosol can, may be used to make any shiny surface slightly matt. One particular type, marketed under the proprietary brand name of 'Anti-flare' may be used on any type of surface and subsequently rubbed off. It does not kill all the reflections completely but gives them a softer character. It is said to be harmless to all surfaces. Nevertheless, it is perhaps better not applied to a priceless painting, just in case.

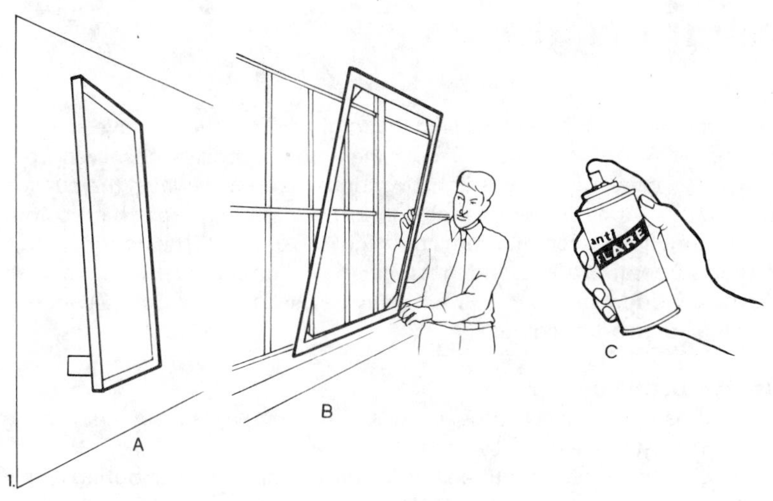


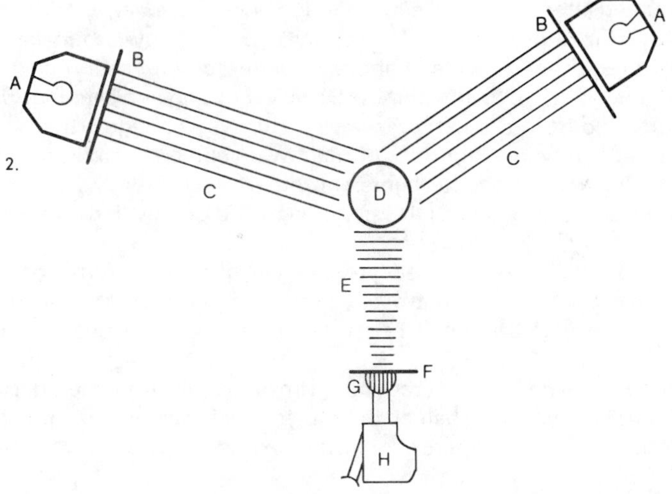


MEANS OF AVOIDING REFLECTIONS

1: A. Picture tipped up by matchbox; B. Window filters set at an angle; C. 'Anti-flare' aerosol used to reduce flare spots. 2: A. Lamps; B. Polarising screens; C. Polarised beams of light; D. Reflective subject; E. All light reflected from subject is polarised in one direction; F. Pola screen in front of camera; G. Light passing to camera excludes reflections; H. Film camera.

*Shooting a night effect by daylight.*

# Day for Night

Theatrical tradition has always held that night effects should have an overall bluish look, no doubt because originally the night was observed from rooms lit by candle, oil or gas light, by comparison with which the outside looked blue. Today, to people looking at the night sky from high colour temperature flourescent-lit rooms the night looks black. The notion of blue moonlight seems artificial and theatrical. If no illuminated area is included in a scene (windows etc.) B & W film may be used to artistic and economic advantage instead of colour.

### Filming practice

Underexposing by two stops destroys shadow detail sufficiently for laboratories to make a heavy print.

If a slight blue tinge is required, for old times' sake, when shooting with a 3200K colour film by daylight, an 81EF filter may be used in place of the 85B or 85, with a CC20G if necessary, to help the faces. With B & W film a blue cast may be introduced in the printing.

Sky and any other bright areas should be excluded as much as possible from the scene and what remains may be darkened down by the use of an ND graduated filter or polar screen. This technique, however, eliminates the possibility of in-shot panning. Roads may be wetted down to make them look darker and their reflected light react better to polarisation.

The 'magic hour' or 'golden minutes' every morning and evening, when, for a short period, there is just sufficient light to give an image in the shadow areas, is frequently used for dusk or night wide angle shots. Street lights, lights in windows, motor car lights and other practical lights which lack intensity show up well during this period and add greatly to the realism of the scene.

As with bright sky, a graduated filter may be introduced from the side or the bottom of the frame to control any excessively large bright areas of the picture. A large black drape may be erected to cast a shadow over a selected area.

The direction of any light source should be principally from the side, three quarters back or even from behind the subject. Night scenes which are flatly lit and lacking in contrast merely look underexposed, murky and characterless.

DAY FOR NIGHT

Night effects may be achieved by: Underexposing by two stops; shooting wide aperture shots at dusk when illuminated windows, signs and car head lights stand out; using graduated filters to darken the sky and foreground; a pola screen may sometimes be used as an alternative means of darkening certain areas.

# Intra-unit Communication

Even at comparatively close range it is sometimes necessary to amplify the human voice by means of a loud hailer or bull horn in order to communicate satisfactorily. A base station in the production office is particularly useful for keeping in touch with the shooting unit, in case of an injury, for relaying rushes (dailies) reports, information regarding the unit wrapping up or moving location, artists availability and as a booster between widely separated units.

## Radio communication

The use of a radio transmitter of any kind is subject to control in almost every country in the world. A licence may be required, the frequency to be used will be stipulated and the power and band-spread on the transmitter subject to restrictions. In the United States and some other countries, 5W 'walkie-talkies' which transmit the so-called 'citizens waveband' (27MHz), may be used without prior authority. But elsewhere, including Britain, not only is 27MHz allocated to other uses (fire, ambulance services in Britain) but a licence is always required. R/Ts may be 'AM' or 'FM' (the two are not compatible) and VHF (30-300MHz) or UHF (300-3000MHz). FM/UHF gives the best results (1MHz = 300 metres). The frequency of an R/T is usually very accurately controlled by a quartz crystal which must be changed whenever a different frequency is required. Crystals must usually be made to order, sometimes taking several weeks. The distance apart at which a pair of R/Ts can communicate clearly and reliably depends upon their power, the local terrain (better with a clear line of sight), and their design characteristics.

Radio microphones and small hand set walkie-talkies usually have a power of 0.5W, portable R/Ts 2–5W, and mains operated base stations, 25W. The range of a fixed position set may be considerably increased by sighting the aerial (antennae) in a high, unobstructed position. The actual length of the aerial is critical and should be whole, half or a quarter of the wavelength in use. Most R/T failures are due to low battery voltage, damaged aerials, electronic failure or an alien environment.

## Radio procedures

When using R/Ts for inter-unit communications on location as well as for communication with the production office it is advantageous to have alternative channels.

All messages, without exception, should begin by saying who is calling whom, and end with the word 'over' if a reply is expected, 'out' at the end of a conversation and 'off' when closing down.

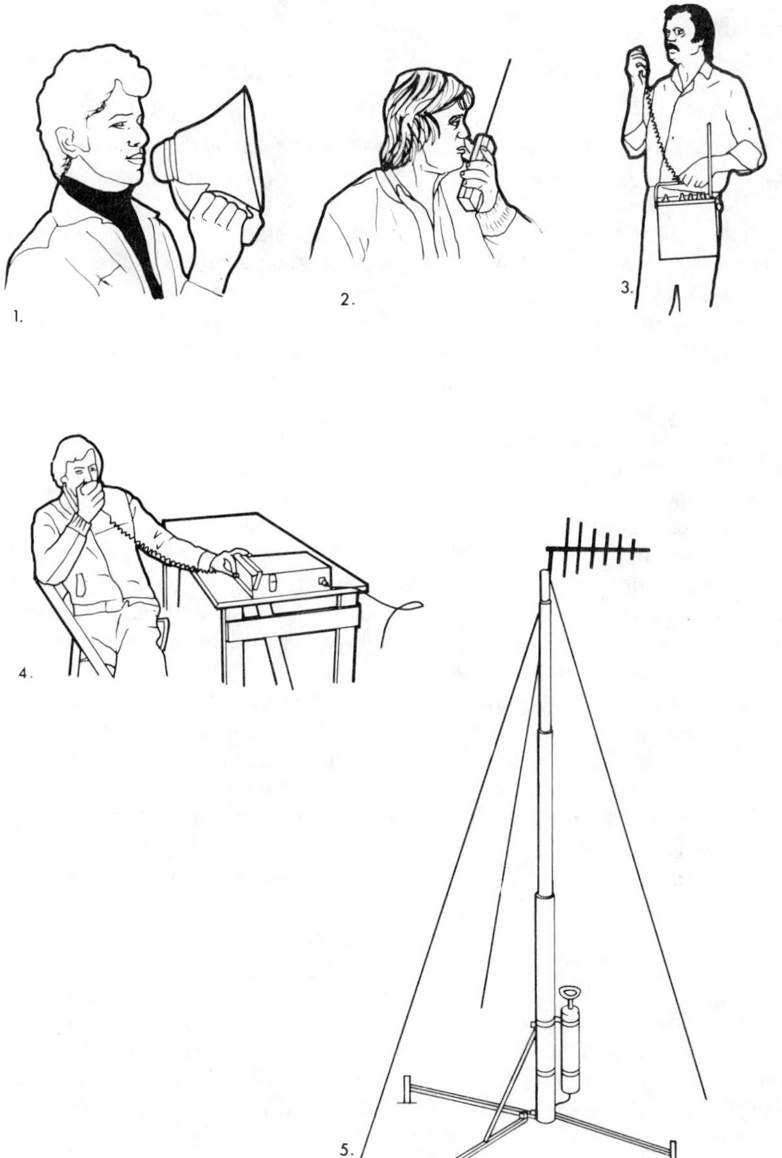

4.

5.

INTRA-UNIT COMMUNICATION

1. Loud hailer or bullhorn; 2. Walkie-talkie; 3. Radio-telephone; 4. Base-station transmitter/receiver; 5. Pump-up aerial.

# Elevation

For many shots a camera position higher than ground level is needed. Elevations of 6 to 13ft (2 to 4m) are usually achieved by the use of a rostrum or parallels. For higher position camera cranes, the roofs of camera cars, scaffolding towers or 'cherrypicker' vehicles are used.

When the roof of a camera car is used for a static set-up particularly if a long focal length is used on the camera, the vehicle springs can be inhibited with jacks, blocks or boxes to ensure a steady platform.

Large, counterweighted camera cranes or boom arms are available, capable of giving camera lens highlights of up to 27ft (8m) above ground level. (See page 134 of *Equipment, Choice and Technique*.)

## Scaffolding towers

Scaffolding towers of prefabricated sections which may be built up to 100ft (30m) should be securely braced and guy-roped if the height is excessive and there is any likelihood of wind. A safety rail should be constructed around the top. With long focus lenses, it may be preferable to fix the camera mount directly to the scaffolding tubes or safety rail rather than simply set a tripod on the floorboards. A high hat with scaffold clamps attached, or a paddle mount, may be used for this purpose. (See page 126 *Equipment, Choice and Technique*.)

When raising or lowering equipment on a high tower, a rope pulley should be used in the proper manner, taking care that the rope is passed underneath equipment for security, the handle(s) being used only to prevent the rope from slipping. The free end of the rope should be wound 2–3 times around a tube to provide a brake to stop a heavy load of equipment slipping downwards out of control. When handling the rope it must be passed from hand to hand and never allowed to slip through when lowering. Many mountaineers' techniques with ropes may usefully be applied and their Karibinders or snap rings are handy accessories for securing a rope and for controlling a load as it is raised or lowered.

## Vehicular elevation

Cherrypickers, which may be extended up to 100ft (30m) are also particularly useful when filming public events. They can place a camera in a commanding position, far above the crowd, low buildings or other such obstructions. The arm may also be extended to place a camera over an obstacle or under a bridge.

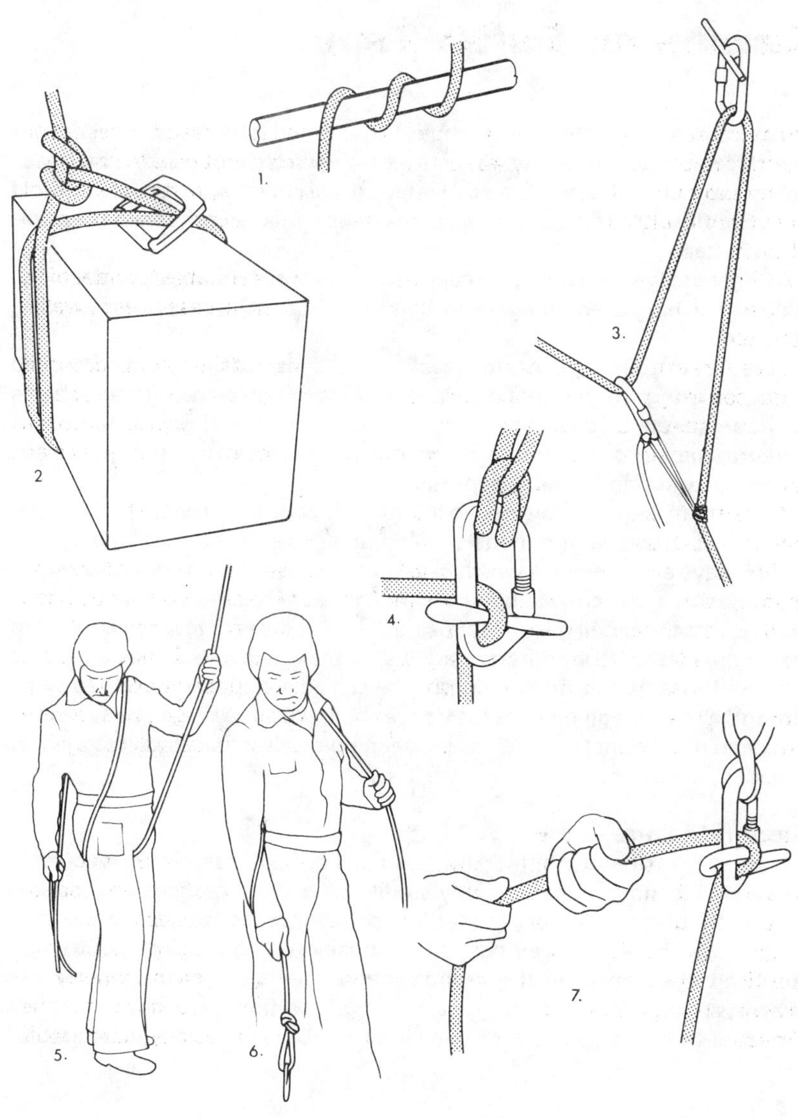


ROPE TRICKS

1: Rope passed three times round a horizontal pole to give extra friction; 2. Correct way to tie a rope around a case; 3. Two Karibinders used to make a pully to lift a heavy weight; 4. Two Karibinders used to apply friction to a rope; 5. Means of lowering a person by himself; 6. Correct way to lower a heavy weight; 7. Ropes should be passed hand over hand and never allowed to slip.

# Camera or Gaffer Tape

Most 'camera' or 'gaffer' tape is about 1in wide and cloth-based. Plastic tape which cannot be torn easily, thin plastic tape which is not pliable and does not mould to the shape of the can edges, transparent tape, the free end of which is difficult to find and cheap cotton tape with little adhesive power are all unsuitable.

A 2in wide type which has great adhesive power is suitable for attaching almost anything to anything, even lightweight lighting accessories to walls, flats, etc.

Uses for camera tape include: sealing film cans containing unprocessed stock, securing a number of cans together, attaching camera report sheets or dope sheets to cans of exposed film, to give a writing surface on a camera, clapper board or film can, to make a distance mark on a lens as an aid to focusing and similar applications.

If a camera is to be subjected to intense buffeting, as on a racing car, tape may be used to seal the magazine and prevent accidental damage.

The ways and means in which camera tape is used is limited only by the imagination of the crew and the circumstances of the situation. It has been used to attach a small mirror alongside the focus scale or footage counter to make it readable from different angles, to make pencil and chalk holders attached to a magazine housing or the rear of a clapper board, to wrap around a lens to pull on for a fast focus change or as a bridge between the lens and the mount to act as a stop when doing a fast focus pull to a given position, and so on.

### General usage

Camera tape is handy for marking positions on the floor to show where an artist should stand, where a dolly should be moved to and where certain distances from the camera are for focusing purposes. Its uses for holding things together, for repairing and for positioning are legion. But when applying tape, especially the strong variety, to someone else's property, as when working on location, great care must be taken to ensure that the surface is not damaged after removal. Wallpaper is especially vulnerable.

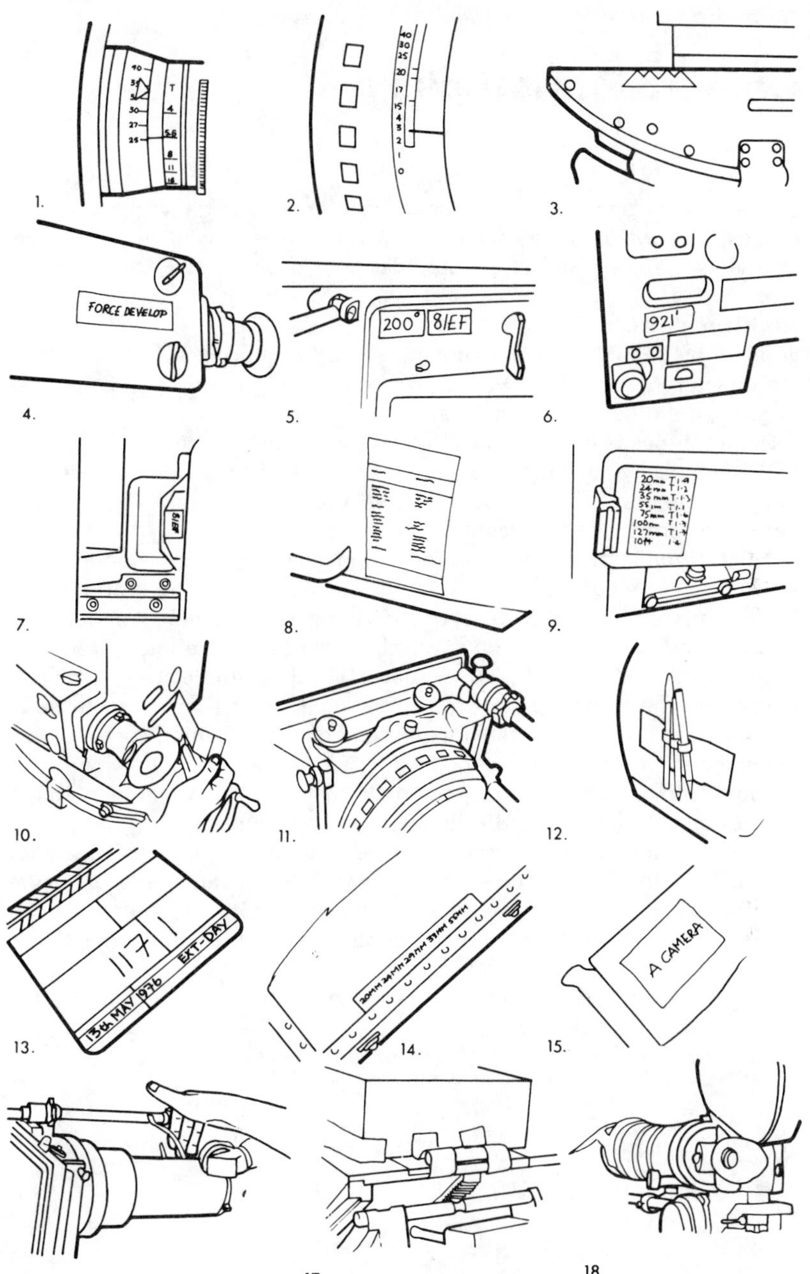


To indicate: 1, zoom; 2, focus; 3, balanced camera position; 4, 5, exposure; 6, film load; and 7, gelatine filter reminders. To attach: 8, call sheet; 9, lens list; 10, mirror to read footage from side; 11, to cut stray light; 12, attach pencils; 13, removable writing surface on clapper board; 14, to mark contents on case; 15, identify hardware; 16, fast focus or zoom stop; 17, attach a level; 18, make a long-lens sight.

# Camera Filmstocks

A cameraman's choice of camera filmstock may be governed by availability, gauge, speed, latitude, whether it is negative or reversal, processing considerations and personal preferences based on experience, chauvinism, prior satisfaction, prejudice and/or price.

## Making a choice

Judged by the final screen image, stocks produced by the principal manufacturers sometimes vary quite considerably in film speed, fineness of grain, definition, colour saturation and image contrast.

Different makes of filmstock may have slight differences in the density of the film base and in the three colour layers. This can be ompensated for in the printing but a cameraman should know the laboratory norm for a particular filmstock and accept it as correct.

All camera filmstocks are normally supplied wound 'emulsion-in', on a standard centre core or on a daylight loading spool (where applicable) with standard negative perforations. They have an identification letter or number and sequential footage numbers exposed as a latent image at regular intervals along the edge. All require the same colour correction filters with the same loss in exposure and all suffer from an increase in grain size, fog level, contrast, imbalance of colour and image degradation when force-processed (see page 28).

Low contrast colour reversal camera filmstocks are available in 16mm and Super 8 gauges and are intended for duplication rather than projection. High contrast 'amateur' reversal films are not satisfactory for duplication. In general, reversal camera originals have less exposure latitude than negatives and without special laboratory attention the faster types may not be suitable for inter-cutting with the slower fine grain types, due to the differences in gamma (contrast) characteristics.

High speed reversal stocks have a very short processing time and some may be force-developed up to a speed of 600–1000+ASA and are particularly well suited to TV news usage.

TECHNICAL CONSIDERATIONS IN CHOOSING CAMERA FILMSTOCKS

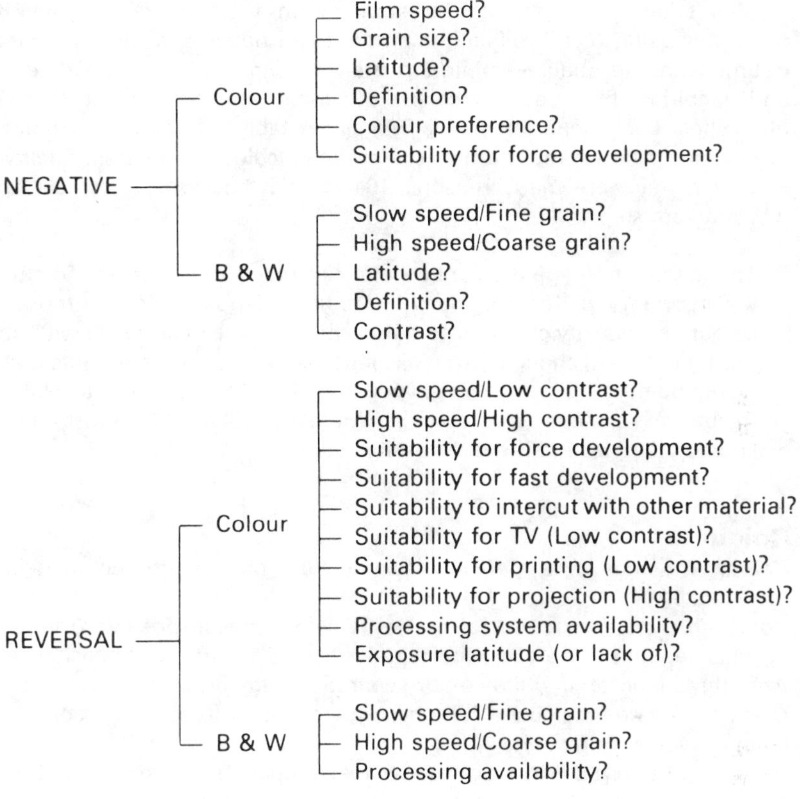

# Duplicating Films

Ideally, show prints should be struck from the camera original, if necessary using A & B rolls in the printer to introduce opticals, to ensure optimum image quality—minimum grain, maximum image steadiness, and definition, best possible colour and tonal reproduction and fewest blemishes. Each generation of printing inevitably introduces definition losses and makes it more difficult to maintain colour or contrast fidelity. However, there are many reasons that make it necessary to create an intermediate step viz:

> To intercut an image with optical effects already incorporated. To alter the geometry of the original for printing purposes. To incorporate colour and density changes so that further prints can be made without grading (timing) changes. To make it possible to strike more prints than would be possible from a single negative. To make negatives available for printing at other laboratories. Where possible, reduction printing should take place at the last possible stage.

## Colour

*Colour Reversal Intermediate*: a one-step negative-to-negative intermediate.
*Colour Intermediate*: used for making a colour master positive from an original negative and a duplicate negative from a colour master positive or from three black and white colour separation negatives.
*Colour Internegative*: used for making a negative from a low contrast colour reversal original.
*B & W separation*: B & W stock used to make separate blue, green, and red duplicate positives from which colour duplicate negatives may be made on colour intermediate stock or to make a B & W master positive from a colour negative, from which, in turn, B & W duplicate negatives may be made for B & W printing.

## Black and white

*B & W duplicating positive*: used to make master positives from B & W negatives.
*B & W duplicating negative*: used to make B & W duplicate negatives from colour or B & W positives.
*B & W high contrast*: used to shoot title artwork, to make mattes for optical bipack printing.
*B & W soundtrack*: film of special gamma used for sound track negatives.
*B & W TV recording*: film of special sensitivity especially suited to photographing the phospor of a B & W TV tube image.

Note Eastman 35mm stock is prefixed '52' and 16mm stock '72'.

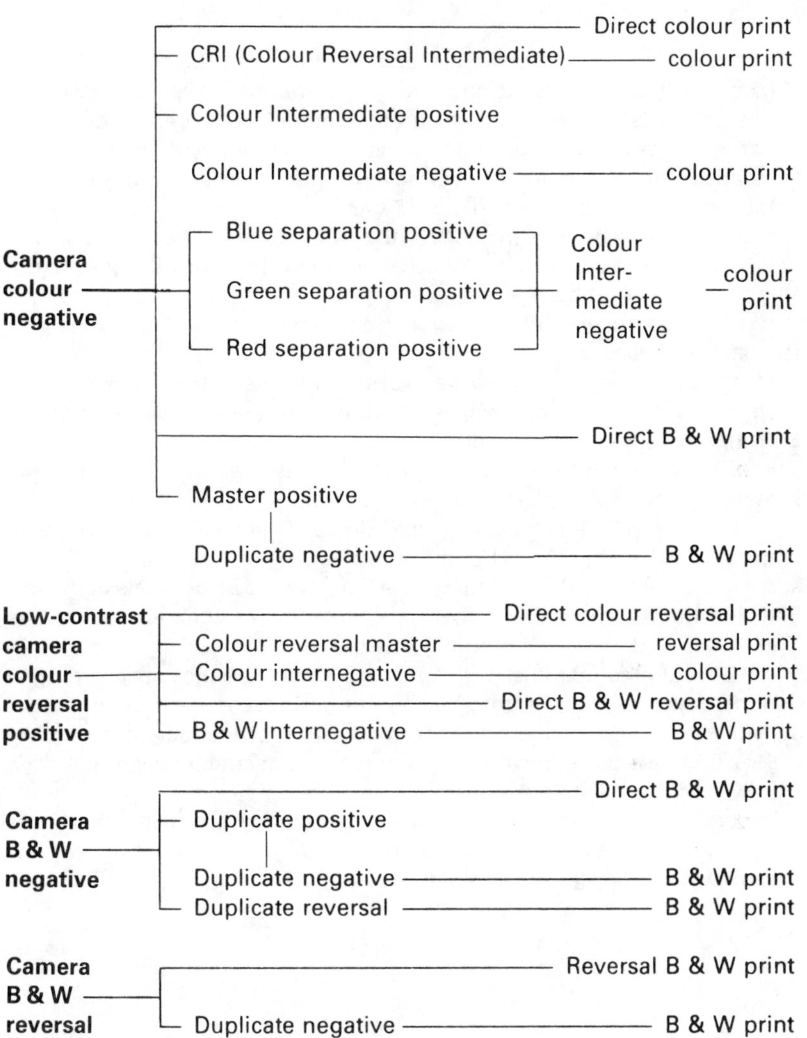
Original
Lab. Process Stock
Result
Direct colour print
CRI (Colour Reversal Intermediate)
colour print
Colour Intermediate positive
Colour Intermediate negative
colour print
Blue separation positive
Camera colour negative
Green separation positive
Colour Inter-mediate negative
colour print
Red separation positive
Direct B & W print
Master positive
Duplicate negative
B & W print
Low-contrast camera colour reversal positive
Direct colour reversal print
Colour reversal master
reversal print
Colour internegative
colour print
Direct B & W reversal print
B & W Internegative
B & W print
Direct B & W print
Camera B & W negative
Duplicate positive
Duplicate negative
B & W print
Duplicate reversal
B & W print
Camera B & W reversal
Reversal B & W print
Duplicate negative
B & W print

# Print Films

Although it happens all too rarely that a cameraman is able to influence the choice of filmstock used to print his film, some knowledge of what is available can assist him in taking a greater interest in the end product. This is particularly important as some stocks are more grainy than others (and thus unable to define fine detail), have greater or less colour saturation and/or greater or less contrast.

Print films may be chosen because they are particularly compatible with the originals or duplicates from which they are to be printed. Or they may be selected because of their colour, tonal or granular characteristics, suitability for use with particular printing equipment and/or processing available or for economic reasons.

Most colour print films may be processed in the Eastman colour print bath. For more rapid processing and a greater throughput stock using the ECP II process may be chosen.

For TV use, and where a low contrast effect is desirable on the cinema screen, TV contrast print stock may be used.

Colour reversal print stocks especially, must be chosen to be compatible with the camera original stock or the type of process available, ie: a different stock is recommended for printing from Ektachrome EF originals than from Ektachrome Commercial and there are other stocks compatible with the Kodak VNF and the Gevachrome II processes.

As with colour print films, B & W print stocks may be chosen according to contrast (such as low contrast for TV) and also according to their colour sensitivity. Blue sensitive orthochromatic film is adequate for printing B & W originals but panchromatic film is more desirable when making a B & W print from a colour negative.

Reversal print stocks are preferable when printing from positive originals as no intermediate negative is needed and a printing generation, with its inherent quality losses, is eliminated.

| *Origin Possibilities* | *Requirement* | *Considerations* |
| --- | --- | --- |
| | | Neg/pos or reversal? |
| | | Grain size or definition? |
| | | Colour saturation? |
| Colour original negative | | Colour cast? |
| Duplicate negative | COLOUR | Colour compatibility with original? |
| Reversal positive | PRINT | Contrast — for projection? |
| Duplicate positive | | — for TV transmission? |
| | | Cost for No. of prints required? |
| | | Processing system availability? |

| *Origin Possibilities* | *Requirement* | *Considerations* |
| --- | --- | --- |
| | | Neg/pos or reversal? |
| B & W negative | | Grain size and definition? |
| B & W reversal positive | B & W | Panchromatic or orthochromatic? |
| B & W duplicate positive | PRINT | Contrast — for projection? |
| Colour negative | | — for TV transmission? |
| Colour negative | | — for Titles? |
| | | — for sound reproduction? |

# Perf, Pitch, Core, Spool, Winding

Filmstocks are manufactured according to rigidly maintained standards.

**Perforations**
All 35mm camera filmstock is perforated with BH type holes. Different holes, types KS, DH (similar to KS but greater corner radii) and CS (narrow, to permit stereo sound tracks) are used for positive stocks.

On 16mm film the same type of perforation is used for both camera and print stocks but may be perforated on both sides (silent) or on one side only (sound) and designated 2R and 1R respectively.

**Pitches**
A number following the perforation type designates the distance from the leading edge of one hole to the similar edge of the next (pitch). The pitch of 35mm negative is 0.1866 in (4.740mm) and positive 0.1870 (4.750mm).

The norm for 16mm camera pitch is 0.2994 in (7.605mm) but for certain high speed cameras, and for 16mm printed positives it is 0.3000 in (7.620mm).

**Cores**
The 200, 400, and 1000ft (61, 122 and 305m) rolls of 35mm negative are usually supplied on 2in type 'U' cores. The 200ft (61m) rolls may also be obtained wound on 1in type 'A' cores. The 200 and 400ft (61 and 122m) rolls of 16mm stock are supplied on 2in type 'T' and 1200 and 2400ft (366 and 732m) rolls on 3in type 'Z' cores.

**Spools**
Certain types of camera are manufactured to take film wound only on daylight loading spools. With 35mm film, 100ft (30m) rolls are available on 'No. 10' daylight spools. Most 16mm film magazines accept film wound on cores or on daylight loading spools and 16mm film is available on 100, 200, and 400ft (30, 61 and 122m) spools (360ft with magnetic stripe).

**Magazines**
Certain 16mm film may be obtained in 50ft (15m) type magazines for use in magazine-loading cameras.

**Windings**
Film is usually supplied wound film emulsion in, but may be obtained emulsion out to special order. Single perforation 16mm film intended for camera use is supplied with the perfs on the right hand side of the film when viewed with the film coming off the top of the roll towards the body ('B Winding'). Laboratories use 'A Winding' for contact printing and 'B Winding' for optical printers.

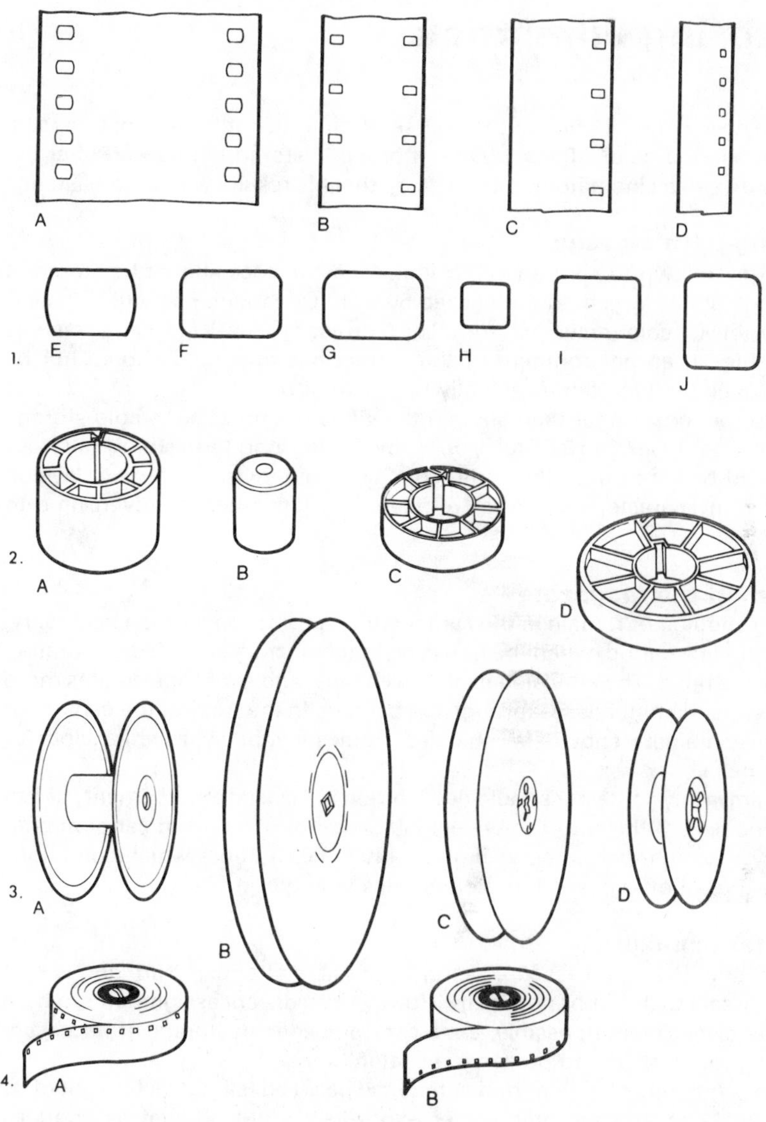


PERF, PITCH, CORE, SPOOL, AND WINDING

1. Perforations: A. 35mm; B. 16mm, double perforations; C. 16mm, single perforations; D. Super 8; E. BH, 35mm negative; F. KS, 35mm; G. DH positive; H. CS; I. 16mm; J. Super 8.
2. Cores: A. 35mm Type R; B. 35mm Type A; C. 16mm Type T; D. 16mm Type Z.
3. Spools: A. 35mm No. 10; B. 16mm 400 ft; C. 200 ft; D. 100 ft.
4. Windings.

# Storing Filmstock

Filmstock is perishable. In the course of time it shrinks and loses speed, contrast and colour fidelity. Only if properly stored before and after exposure can a cinematographer be sure that his results will be consistent.

### Long-term storage

For periods up to six months, colour negative stock should be stored in temperatures which do not exceed 50°F (10°C), though 65°F (18°C) is permissible for colour reversal stock. The tape used to seal a film can preserves the film in an environment of the correct humidity, and should not be removed until the film is actually required for use.

For periods longer than six months, filmstock must be in cold storage conditions of around 32°F (0°C) or below. During long-term storage all stock should be kept under the same conditions otherwise rolls may develop differing characteristics. After refrigeration, film must be given time to normalise.

### Film in the magazine

Film should not remain in the camera or magazine longer than necessary. During this period, when the film is no longer factory sealed, deterioration is more rapid. The emulsion itself can change, and shrinkage creates more noise as the film passes through the camera. In the tropics, the camera or film containers should be shielded from sunlight by a white cloth or umbrella.

Film in any container should never be left in a closed environment subject to the heat of the sun. The unventilated interior of a parked car or aircraft may reach temperatures as high as 140°F (60°C) under which conditions unexposed or exposed film deteriorates very quickly.

### After exposure

Film should be processed as soon as possible after exposure and is normally sent to the laboratory daily. However, on distant assignments, where daily dispatch is impossible, extra care is needed in storage, if necessary using refrigeration, to preserve the latent image.

In conditions of high humidity the time between taking the film out of its original can and recanning after exposure should be kept as short as possible. Immediately after loading, magazines should be put into new polythene bags and sealed until required. They should be left on the camera for as short a time as possible and returned to a moisture-proof container until unloaded and sealed in the original cans. Desiccants may be used as an aid to keeping film dry.

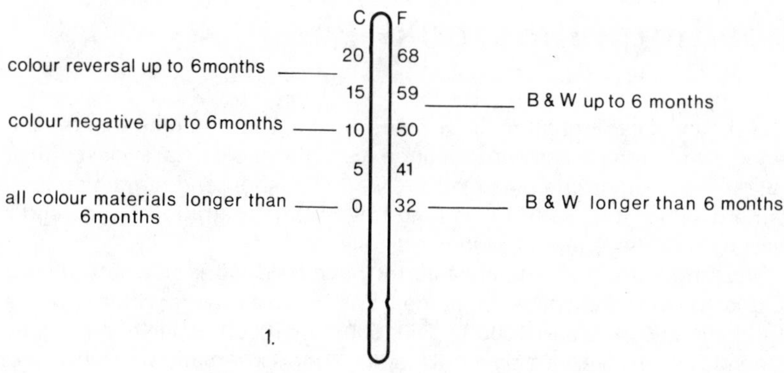


FILM STORAGE

1. Recommended maximum temperatures.

FILM WARM-UP TIMES AFTER COLD STORAGE

| Temperature difference between store & outside °C | °F | outside relative humidity per cent | warm-up time (hours) single roll 16mm | 35mm | carton of 10×35mm |
|---|---|---|---|---|---|
| 14 | 25 | 70 | ½ | 1½ | 12 |
| 14 | 25 | 90 | 1 | 3 | 28 |
| 55 | 100 | 70 | 1 | 3 | 30 |
| 55 | 100 | 90 | 1½ | 5 | 46 |

# Loading Filmstock

The choice of camera often determines the magazine loading that must be worked with, regardless of the economics of the choice. Rarely is it practicable to run a magazine out completely and the short end remaining is the same whether it is a short or a long roll of film. The ratio of film used to wastage is always higher with short rolls.

Short rolls are more practicable for hand-held work or where there is restricted overhead space. Most ideal are 35mm cameras with magazines that have 200, 400 and 1000 or 250, 500 and 1000ft capacities or 16mm cameras which take 100ft daylight loading spools internally and 200 or 400ft magazines externally as required.

## 16mm

For 16mm usage there are different problems. On the one hand there is the need to have a magazine load which runs for as long as possible, especially in the case of TV news and documentary work where sound is to be recorded. On the other hand, there should always be a spare magazine if for no other reason than the one on the camera may jam. A reasonable compromise is to use 200ft daylight loading spools which run for 5½ min in 'clip-on' magazines. Spare rolls may be carried in pockets without the need to carry extra magazines in a heavy magazine case.

## Loading magazines

A darkroom in which to load and unload magazines is a luxury. All too often a changing bag must be used. The changing bag must be free of dust. Small particles picked up from the changing bag material may become hairs in the gate and sparkle on the screen. A changing bag should be used in a shaded area, never in direct sunlight. The loader's hands should be pushed well into the arm holes to ensure that no light passes through.

## Daylight loading cameras

To avoid spoiling the first and last shots on a roll it is important never to load or unload daylight spools in a bright light. If it is known that important material is on the very end of a scene it is possible to use a coat as an emergency 'changing bag' by wrapping the body of the jacket around the camera and putting the hands in through the sleeves in reverse.

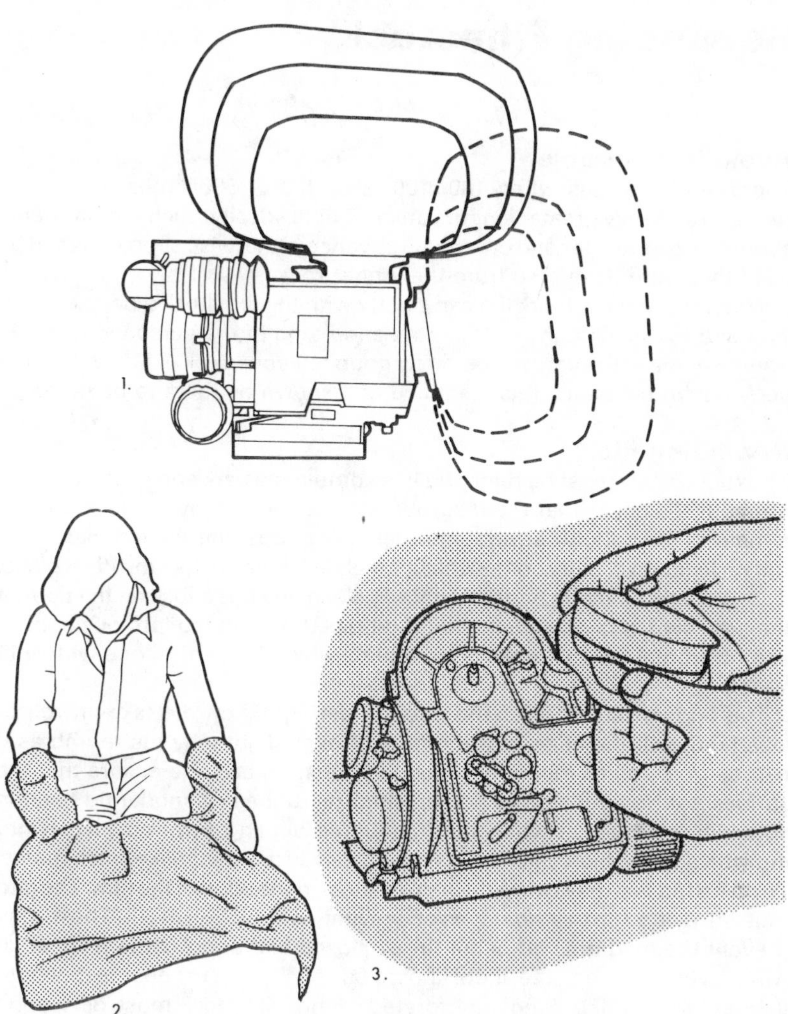


FILM LOADING

1. Panaflex camera showing alternative magazine sizes and positions; 2. Reloading a magazine in a changing bag, an operation which should be done in the shade; 3. Removing the light protection band from 100 ft daylight loading spool. This should be done in as dark an area as is possible in the circumstances.

# Rewinding Filmstock

## Avoid it if possible

There are occasions when 100, 200, 250, 400 or 500ft rolls of film are required and only greater lengths are available. Ideally, such requirements should be planned for and ordered in advance. Otherwise short ends can be used that were generated from the longer rolls.

When re-canning, mark the can clearly with the filmstock type, the emulsion and the batch number, the date and name of the person who did it. Some people will not trust a perfectly good re-canned roll of film unless the person who re-canned it is identified and known by them to be reliable.

## Rewinding film

Unexposed film must be removed in a completely dark and dust-free environment. To check that the blackout is complete, it may be necessary to remain in the dark for several minutes before opening the film can.

If on location, a bathroom within a hotel bedroom is often the ideal place. It can usually be made perfectly dark, is clean and has a lock on the door. If the light switch is outside it should be taped over in the 'off' position.

A horizontal flat plate rewinder is preferable to a vertical one with split spools.

The fingers on the left hand should rest lightly on the take-off roll to steady it, with the edge of the film bent over slightly by the thumb as it passes from roll to roll. Very great care must be taken to ensure that the surface of the emulsion is not touched or rubbed. A moderate tension should be maintained evenly throughout the winding. Erratic or jerky winding with sudden increases in tension will produce scratch-like cinch marks. Winding too fast may cause flashes of static on the edge of the film. The film must be wound so that no ridges of unevenly wound film occur. These may prevent the magazine lid being firmly closed and affect camera speed. If winding onto a spool, the spool should be checked first to ensure that both flanges are parallel. Single-perforated 16mm filmstock must be wound twice to ensure that the winding remains correctly handed, otherwise it cannot be threaded into the camera in the right way. If winding to a certain length, a piece of stick, cut to the diameter of the required roll may be used to measure film in the dark.

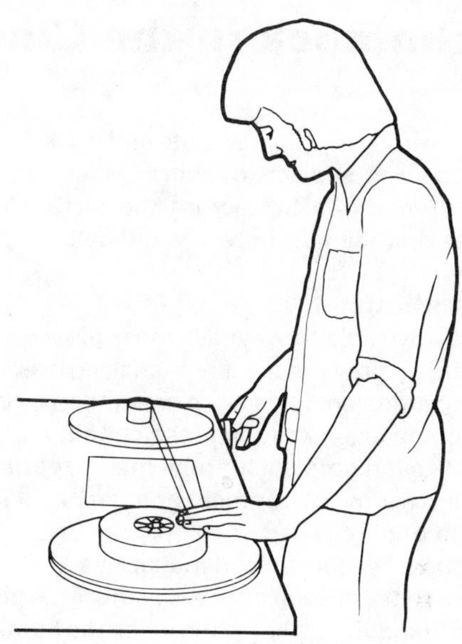

1. Horizontal rewinder; it produces even windings with the minimum difficulty.

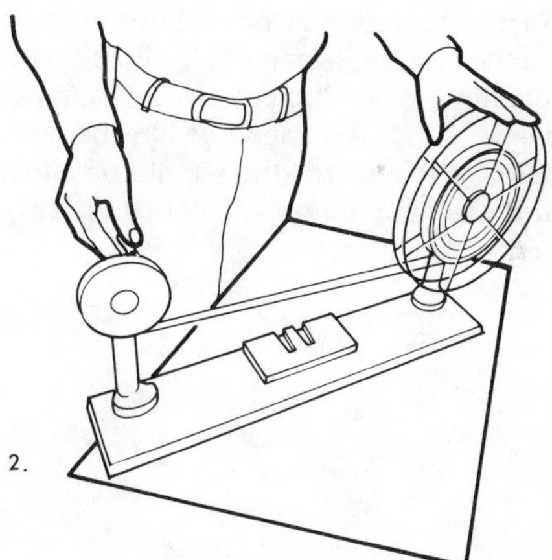

2. Vertical rewinder; it is not so easy to use properly but is more portable.

# Filmstock in the Cold

Below freezing point acetate filmstock dries out and becomes progressively brittle and subject to 'static' which, on the screen, looks like flashes of lightning streaking across the picture area. Loading a camera without breaking the film becomes difficult.

### Handling

It is advisable to wear silk inner gloves when loading extremely cold film, both to prevent the flesh sticking to any bare metal surface and as a precaution against fingers being deeply cut by the edges of the film, which become like a knife in these conditions.

To overcome such problems, magazines may be loaded in a comparatively warm environment and then kept in a thermally insulated magazine case until required.

It is advisable to load magazines as shortly before use as possible and to unload them soon after exposure. This gives the advantage of confining all film handling to the period when the humidity is near normal—as preserved by the hermetic seal given to the film when originally packaged by the manufacturer.

Many loading problems may be alleviated by choosing a camera with magazines which incorporate the film drive sprocket, the rollers, the film guides and preformed loops. Such magazines then need only to be clipped on to the rear of the camera for loading purposes. With no handling of the actual film involved there is less likelihood of its breaking.

Below −40°F the speed of film begins to drop and between −50°F or −60°F a whole extra stop should be allowed.

At these temperatures, Estar or Mylar base film is preferable to acetate stock.

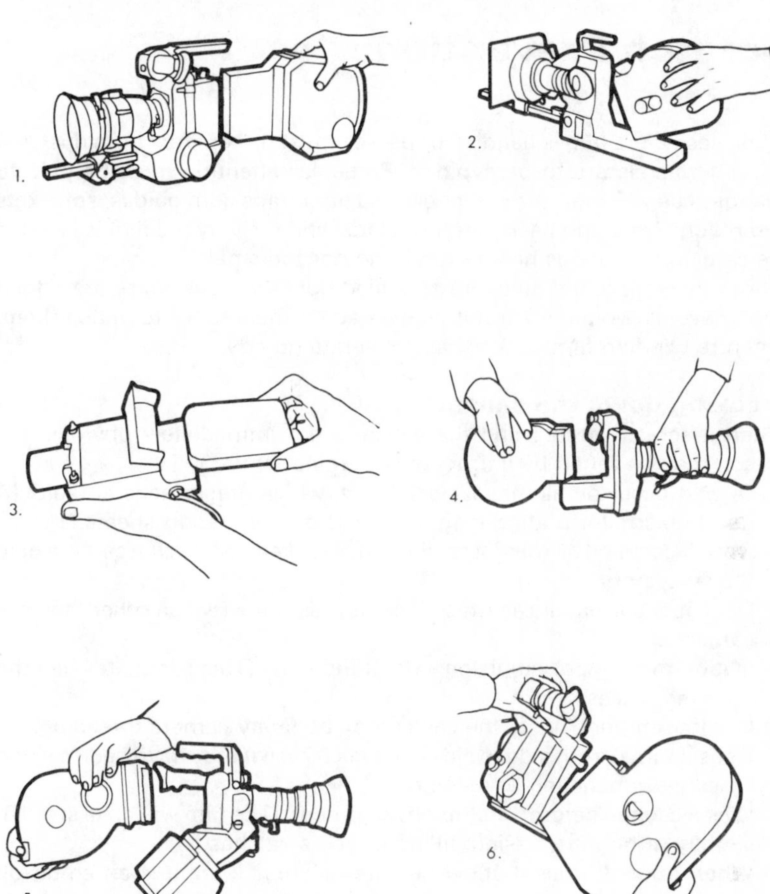


COLD CONDITIONS

Cameras with magazines having preformed film loops which are particularly suitable for use in cold conditions.
1. Aaton 400ft magazine 16mm; 2. Arriflex 16SR 400ft magazine 16mm; 3. Bell & Howell 50ft cartridge 16mm; 4. Eclair ACL 200 or 400ft magazine 16mm; 5. Eclair NPR 400ft magazine 16mm; 6. Eclair CM3 400ft magazine 16mm.

# Scratch Prevention

All places where film is handled or passes must be kept free from dust and grit if scratches are to be avoided. Particular attention must be paid to changing bags, magazine chambers and light traps, film guides, sprockets and rollers, front and back pressure plates and the way the film is laced in the camera with loops neither too large nor too small.

Loosely wound film must not be pulled tight and rolls which are ridged and unevenly wound must not be banged on their edges to flatten them. When re-winding filmstock never accelerate quickly.

## Tracking down the cause

When a scratch occurs and the cause is not immediately obvious, the reason may be found by a process of elimination:

1. Does it occur on all magazines? If not, which magazine is at fault? (A good reason for putting mag. numbers on neg. report sheets.)
2. Does it occur on all rolls? If so, the cause is likely to be in the camera or in the laboratory.
3. Does it occur on all cameras? If so, it must be a reason other than the camera.
4. Is it on the base or emulsion side of the film? (This eliminates half the possible places.)
5. Is it intermittent? If so, the cause may be faulty camera threading.
6. Does it move from side to side—in which case it may have been caused by careless handling or re-winding.
7. Was it caused before, or after development. (The film will have swollen slightly during processing, filling in fine scratches.)
8. When printed, does it show definite colours? If so, it is an emulsion scratch.
9. Is it on the track side, centre, or right hand side of the screen? (The track side is on the inside of the camera.)

When the facts are established, a roll of virgin filmstock should be carefully loaded into the suspect equipment and the places where it passes through the magazine light traps and the top and bottom of the pressure plate marked with a pencil. The camera should be run and the various points marked again. The film is then unthreaded and the scratch looked for. By observing where the scratch starts and finishes in relation to the marks it is possible to pinpoint where the scratch occurs. Scratches caused by grit may possibly be cured by blowing, brushing or wiping. Scratches caused by a pressure plate or film guide becoming abrasive must be cured by rubbing the offending surface with an orange stick, or soft plastic and nothing more abrasive than jeweller's rouge.

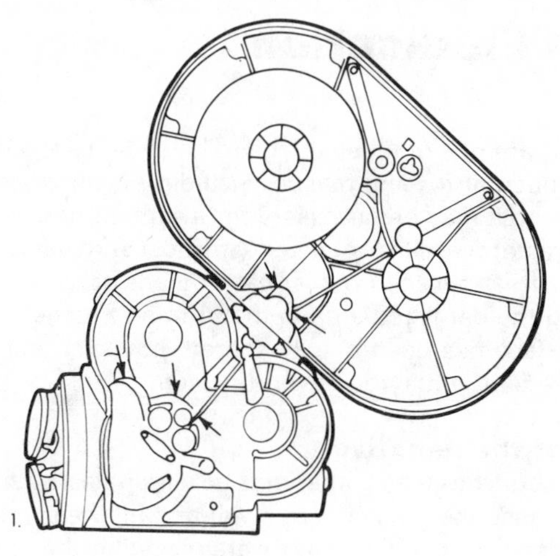

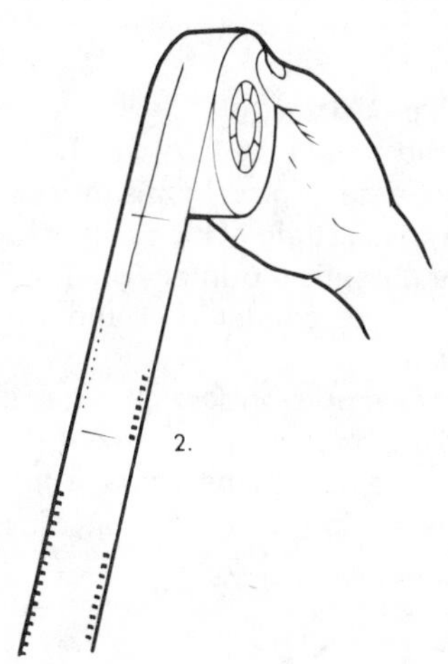

TRACKING DOWN THE SOURCE OF SCRATCHES

1. Points in a 16mm Arriflex St which might be marked to trace the source of a camera scratch; 2. Examining a reel of 35mm film which has been scratched as it passed through the camera. Note the marks made either side of the gate exactly pinpoint the source.

# Scratch Eradication

While scratches are one of those things to be avoided at all times they do sometimes happen. If for any reason, including economics, the material cannot be re-shot, it may be necessary for a cameraman to explain to the producer or director that all is not necessarily lost and that laboratories and specialised film treatment companies have means of eradicating, minimising or avoiding the damaged area in the printing process.

Once the surface has been deeply scored, however, especially on the emulsion of the film, it may be beyond rescue.

## Treatment of the negative

Fine emulsion scratches may sometimes be eliminated by soaking the film sufficiently to cause the emulsion to swell and fill the scratch.

Fine scratches on the cell side may be eradicated by the use of a film base polishing machine. Scratch removal is accomplished by applying controlled amounts of solvent to the film base and reforming the base on a frosted glass wheel. The scratches are then removed by filling-in when the film is re-run over a highly polished glass wheel.

## Treatment by duplication

Scratches on the celluloid side of the film may be filled during the printing stage by the use of a wet-gate printer. In this process the damaged original is totally immersed in a special fluid of the same refractive index as the base while it is in the gate of an optical printer. The liquid fills any scratches or blemishes. Provided that the emulsion is undamaged no scratches will show up on the duplicate.

A 'last resort' cure for an emulsion scratch, especially if it is near the edge of the frame, is to make a duplicate negative (CRI) or print on an optical printer, enlarging and repositioning the frame in order to avoid the scratch.

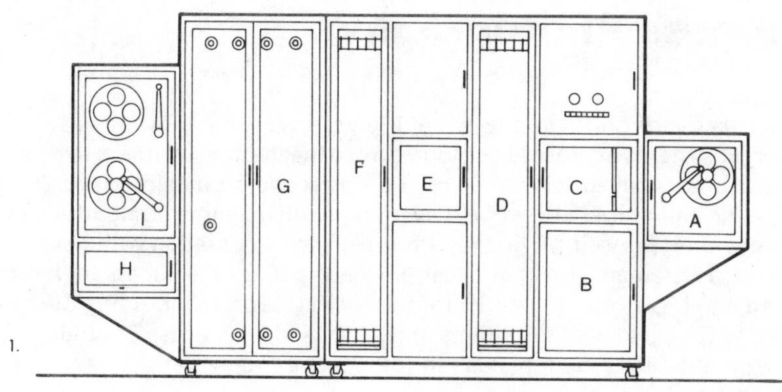


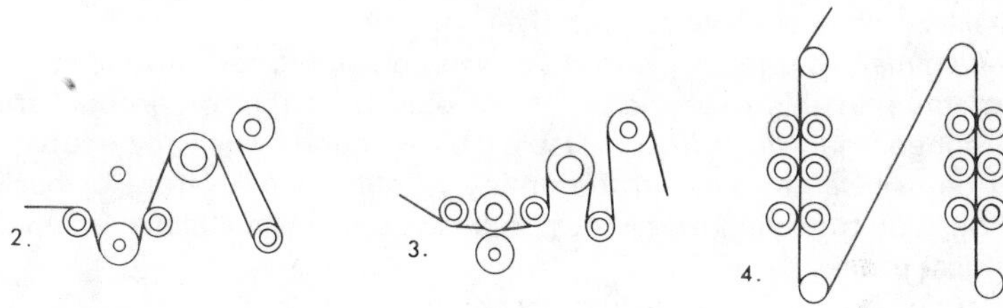


SCRATCH ERADICATION

1. H.F.C. Film rejuvenation machine: A. Taking off reel; B. Pre-cleaning section where all residual dirt is removed by buffing rollers; C. Base scratch removal. Film is immersed in a filling solution as it passes between stainless steel rollers; D. Drying section; E. Emulsion scratch removal. An aqueous base solution is applied which swells the emulsion; F. Second drying section; G. Curing section; H. Lubricator; I. Take-up. 2. Lubricator section; 3. Scratch removal section; 4. Dirt removal section.

# Process Photography

The director of photography should have a working knowledge of the different systems of process photography so that he may use them to create special effects called for in the script or to reduce production costs.

If sufficiently experienced, he may prefer to do the more straightforward process shots himself. Or he may work in close co-operation with a special effects cameraman. On other occasions he shoots all the necessary foreground and background material for the optical laboratory printing department to complete. In the UK films and stills used for process photography backgrounds are called 'plates'. In the US still transparencies are called 'stereos'.

While some aspects of special effects photography are too time-consuming to be undertaken by the main unit, with all the personnel involved, other processes actually same time on the studio floor and may well involve the presence of principal artists. In that case the overall director of photography is responsible.

## On-the-floor methods

In *front and rear projection* a still or motion picture image is projected simultaneously with the action so that foreground action and background scenes are combined photographically.

*Travelling matte or blue backing* is a process of adding background at the printing stage. In this case the artists or objects in the foreground are photographed against an intensely blue background which may subsequently be used as a matte in the optical printing. Unlike front or back projection, with travelling matte the composite scene is not achieved until after optical printing.

*Glass shots*. A scale image accurately painted on a piece of glass, or a cut out still photograph, firmly fixed at an appropriate distance where the depth of field matches that of the background, is used to add to a scene or to replace an unwanted area.

A *scale model* of the upper part of a building or room may be hung in front of the camera and photographed while action takes place below. This saves building a very expensive upper part of a set which may only be used for one brief establishing shot or is used to modify a part of an existing building.

PROCESS PHOTOGRAPHY

1. **Front projection**
A. Scotchlite screen;
B. Foreground subject;
C. Semi-silvered mirror;
D. Background projection;
E. Camera; F. Black baffle.

2. **Back projection**
A. Projector; B. Translucent screen; C. Foreground subject;
D. Camera.

3. **Travelling matte**
A. Blue screen; B. Foreground subject; C. Foreground camera;
D. Background scene;
E. Background camera;
F. Optically combined print.

4. **Glass shot**
A. Background scene of low building; B. Matching glass painting of upper storeys;
C. Camera; D. Combined scene.

5. **Hanging miniatures or scale models**
A. Scene of low building;
B. Model of upper storeys;
C. Camera; D. Combined scene.

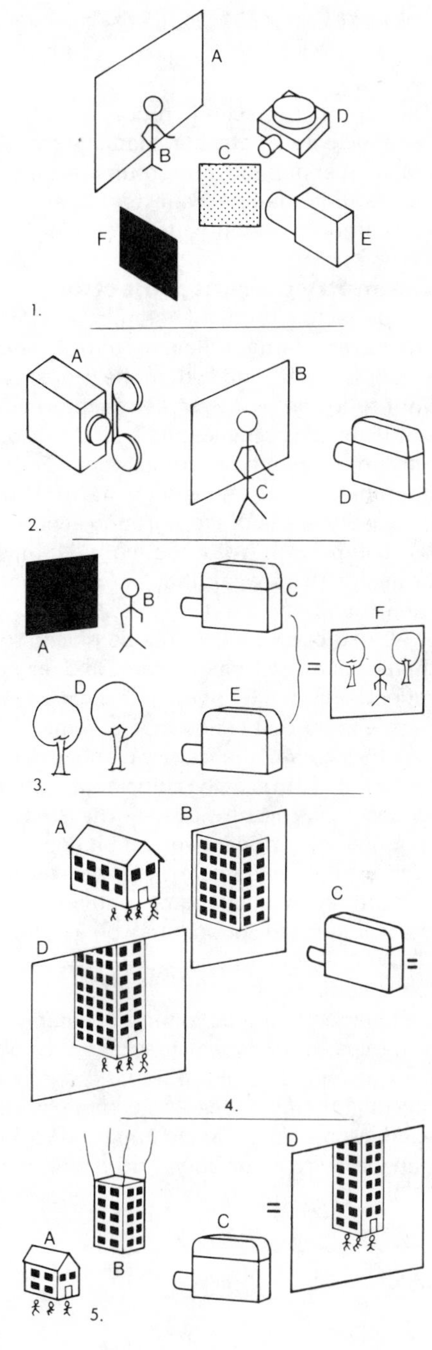

# Front Projection

The front projection process uses a special material (type 7610 hi-reflectivity 'Scotchlite' made by the 3M Company) for a screen which is placed behind the foreground action. Like road signs with tiny glass lenses, this material reflects directionally back to the source almost the total amount of light surface shone onto it.

## Geometry of front projection

By projecting either a stereo or cine 'plate' through a semi-silvered (spattered) partial reflecting mirror, placed in front of the camera lens at an angle of 45°, and setting the projector at right angles to the camera, the light reflected by the screen is returned directly into the camera lens.

It is usual for the semi-silvered mirror to reflect 50 per cent of the light and to transmit the remainder. Thus, sufficient light is reflected to illuminate the screen and the mirror is transparent enough for the camera to be able to see the foreground action and the reflection from the screen. By comparison with the light illuminating the scene, that from the projector is so weak that the projection beam will not show up on the foreground, even if it is white.

A dead black baffle must be placed to the side of the camera, opposite the projector, so that the camera does not also see an unwanted image as reflected by the 50 per cent reflectivity of the glass.

To ensure that solid objects in the foreground hide their own shadows on the screen, the camera and projector and partial mirror must be mounted with exactly coincident optical axes. The front pupils of both lenses must occupy exactly the same point in space. If the camera is mounted on a nodal pan and tilt head so that it moves about the entrance pupil of the lens, in-shot pan and tilt camera movements may be made without any shadowing or fringing. If a zoom lens is used on the camera the front entrance pupil may be just two, or three inches behind the front element at wide angle and be six inches behind the film plane at long focus settings.

Plates and stereos should be made on the largest format possible to minimise grain, which may show up when only a small portion of the background is re-photographed eg: filming a close-up of a foreground object. For cine plates, projectors with full (silent) apertures are normally used. Movement may be imparted to stereo backgrounds by moving the transparency up or down, from side to side and/or rotating it.

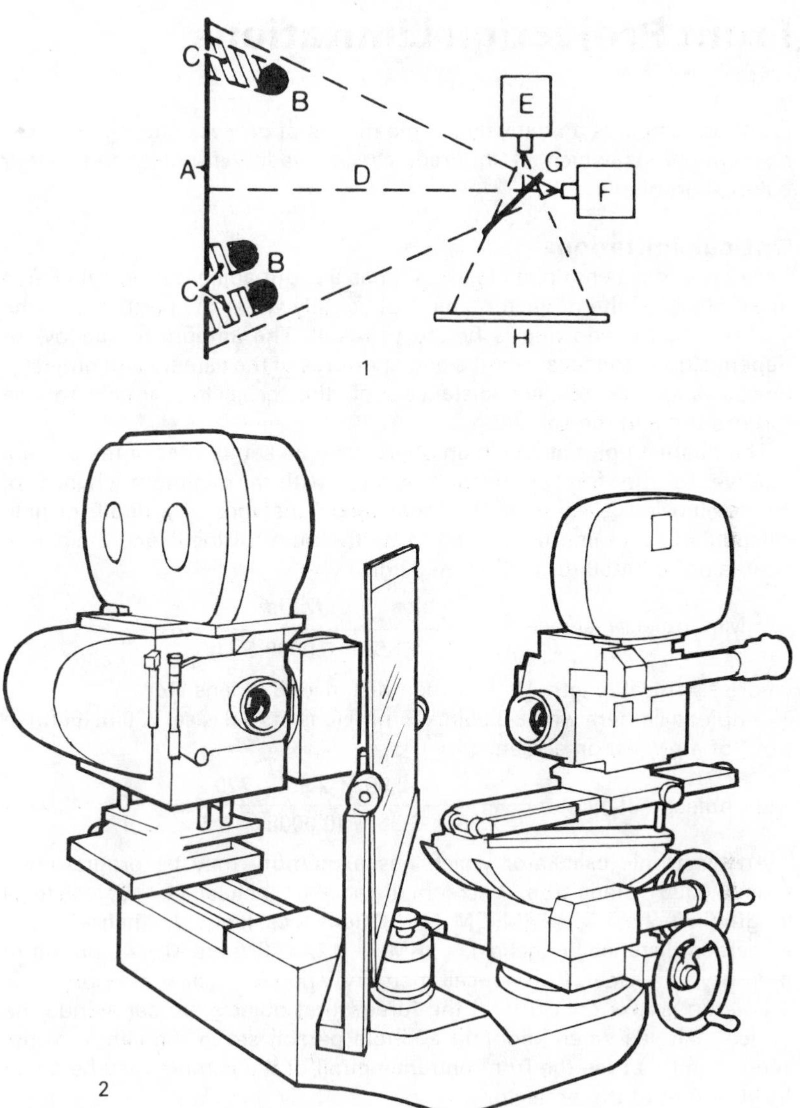


FRONT PROJECTION

1. Principle of front projection layout: A. Scotchlite screen; B. Foreground; C. Foreground shadows; D. Optical axis; E. Projector; F. Camera; G. Semi-reflecting mirror; H. Black absorber screen.
2. Typical front projection systems.

# Front Projection Limitations

Front projection is a relatively simple means of process photography but has limitations which, if ignored, cause shadow fringing and colour balance problems.

## Optical limitations

Because a lens is not pinhole sized it has a slight ability to see either side of an object. If the projector lens is physically wider in aperture than the camera lens, shadowing is bound to result. The amount of shadowing depends upon the focal lengths and apertures of the camera and projector lenses and the relative distances of the projector/camera to the foreground and the screen.

The nearest point at which an object may be set in front of the camera relative to the front projection screen with a minimum chance of shadowing is the same as the 'near focus' distance of a depth of field calculation assuming the screen to be the point of focus and a circle of confusion of 1/250 (0.0008) in. (0.02mm) viz.

$$\text{Min. object distance} = \frac{f2\ \text{Us}}{f2 + 0.0008\ \text{S Us}}$$

where $f$ = focal length, Us = screen distance, S = lens stop.
Example: a camera with a 2.95in. (75mm) lens at $f$5.6 is set 720in. (60ft.) in front of a projection screen.

$$\text{Min. object distance} = \frac{2.95 \times 2.95 \times 720}{2.95 \times 2.95 + (0.0008 \times 5.6 \times 720)} = 43.77 \text{ ft}$$

An electronic calculator which has a memory may be programmed thus: 0.0008, × lens stop, × screem distance (in inches), M +, C, lens focal length (in inches) ×, =, RM, CM, M+, C, lens focal length (in inches), ×, =, ×, screen distance (in inches), ÷, RM, ÷, 12, =. (Where M+ = commit to memory, C = clear, RM = recall memory, and CM = clear memory.)

Line-up is less critical if all the foreground objects are set astride the optical axis, as when shooting a single person set in the centre of the screen. In this case the front entrance pupil of the camera may be set in front of that of the projector.

## Lighting

Light used for foreground subjects should not fall in the background as it reduces its image contrast. If the set is lit to an intensity of 400ft candles (4300 Lux), 100ft candles (1070 Lux) will be seen by the camera, the remainder being lost to the mirror.

Front projection tends to encourage the use of very large screens because so little light is required, owing to the efficiency of the reflective material. 250 watts of light is sufficient to illuminate a screen 30–40ft (9–12m) wide.

## FRONT PROJECTION LIMITATIONS

MINIMUM OBJECT DISTANCE (Feet & Inches)
(Circle of confusion = 0.0008in.)

| Screen Distance | Lens Focal Length | Lens Aperture 1.4 | 2.0 | 2.8 | 4.0 | 5.6 | 8.0 | 11.0 | 16.0 |
|---|---|---|---|---|---|---|---|---|---|
| 10ft | 25 | 8–9 | 8–4 | 7–10 | 7–2 | 6–5 | 5–7 | 4–9 | 3–10 |
| | 32 | 9–3 | 8–11 | 8–7 | 8–0 | 7–6 | 6–8 | 6–0 | 5–1 |
| | 40 | 9–6 | 9–3 | 9–0 | 8–8 | 8–3 | 7–8 | 7–0 | 6–2 |
| | 50 | 9–8 | 9–6 | 9–4 | 9–1 | 8–9 | 8–4 | 7–10 | 7–2 |
| | 75 | 9–10 | 9–9 | 9–8 | 9–7 | 9–5 | 9–2 | 8–11 | 8–6 |
| | 100 | 9–11 | 9–10 | 9–10 | 9–9 | 9–8 | 9–6 | 9–4 | 9–1 |
| 15ft | 25 | 12–5 | 11–7 | 10–7 | 9–5 | 8–2 | 6–10 | 5–8 | 4–5 |
| | 32 | 13–4 | 12–8 | 11–11 | 11–0 | 9–11 | 8–8 | 7–6 | 6–1 |
| | 40 | 13–10 | 13–5 | 12–10 | 12–2 | 11–4 | 10–3 | 9–2 | 7–9 |
| | 50 | 14–3 | 13–11 | 13–7 | 13–1 | 12–5 | 11–7 | 10–8 | 9–5 |
| | 75 | 14–8 | 14–6 | 14–4 | 14–1 | 13–9 | 13–3 | 12–8 | 11–10 |
| | 100 | 14–9 | 14–9 | 14–7 | 14–5 | 14–3 | 13–11 | 13–7 | 13–0 |
| 20ft | 25 | 15–8 | 14–4 | 12–10 | 11–2 | 9–6 | 7–9 | 6–3 | 4–9 |
| | 32 | 17–1 | 16–1 | 14–11 | 13–6 | 11–11 | 10–2 | 8–7 | 6–10 |
| | 40 | 18–0 | 17–4 | 16–5 | 15–3 | 13–11 | 12–4 | 10–10 | 8–11 |
| | 50 | 18–8 | 18–10 | 17–7 | 16–8 | 15–8 | 14–4 | 12–11 | 11–2 |
| | 75 | 19–5 | 19–2 | 18–10 | 18–4 | 17–10 | 17–0 | 16–1 | 14–9 |
| | 100 | 19–8 | 19–6 | 19–4 | 19–0 | 18–8 | 18–2 | 17–7 | 16–8 |
| 30ft | 25 | 21–2 | 18–10 | 16–6 | 13–8 | 11–3 | 8–11 | 6–11 | 5–2 |
| | 32 | 23–11 | 22–0 | 19–10 | 17–5 | 14–10 | 12–3 | 10–0 | 7–8 |
| | 40 | 25–10 | 24–4 | 22–8 | 20–6 | 18–2 | 15–7 | 13–2 | 10–6 |
| | 50 | 27–2 | 26–1 | 24–10 | 23–2 | 21–2 | 18–10 | 16–6 | 10–6 |
| | 75 | 28–8 | 28–2 | 27–5 | 26–6 | 25–4 | 23–9 | 22–0 | 19–7 |
| | 100 | 29–2 | 28–11 | 28–6 | 27–11 | 27–11 | 26–1 | 24–11 | 23–1 |
| 45ft | 25 | 26–11 | 23–10 | 20–0 | 16–1 | 12–10 | 9–10 | 7–7 | 5–6 |
| | 32 | 32–7 | 29–2 | 25–6 | 21–7 | 17–10 | 14–2 | 11–3 | 8–5 |
| | 40 | 36–2 | 33–10 | 30–3 | 26–6 | 22–9 | 18–10 | 15–5 | 11–11 |
| | 50 | 38–11 | 36–10 | 34–3 | 31–2 | 27–8 | 23–10 | 20–3 | 16–2 |
| | 75 | 42–1 | 40–11 | 39–6 | 37–7 | 35–3 | 32–3 | 29–1 | 25–1 |
| | 100 | 43–4 | 42–7 | 41–9 | 40–6 | 38–11 | 36–10 | 34–5 | 31–1 |

MINIMUM OBJECT DISTANCE (Metres)
(Circle of confusion = 0.02mm)

| Screen Distance | Lens Focal Length | Lens Aperture 1.4 | 2.0 | 2.8 | 4.0 | 5.6 | 8.0 | 11.0 | 16.0 |
|---|---|---|---|---|---|---|---|---|---|
| 3 m | 25 | 2.65 | 2.52 | 2.37 | 2.17 | 1.95 | 1.70 | 1.46 | 1.18 |
| | 32 | 2.78 | 2.69 | 2.58 | 2.43 | 2.26 | 2.04 | 1.82 | 1.55 |
| | 40 | 2.85 | 2.79 | 2.72 | 2.61 | 2.48 | 2.31 | 2.12 | 1.88 |
| | 50 | 2.90 | 2.86 | 2.81 | 2.74 | 2.65 | 2.52 | 2.37 | 2.17 |
| | 75 | 2.96 | 2.94 | 2.91 | 2.88 | 2.83 | 2.76 | 2.69 | 2.56 |
| | 100 | 2.98 | 2.96 | 2.95 | 2.93 | 2.90 | 2.86 | 2.81 | 2.74 |
| 5 m | 25 | 4.09 | 3.79 | 3.45 | 3.05 | 2.64 | 2.19 | 1.81 | 1.40 |
| | 32 | 4.40 | 4.18 | 3.93 | 3.60 | 3.23 | 2.81 | 2.41 | 1.95 |
| | 40 | 4.60 | 4.44 | 4.26 | 4.00 | 3.70 | 3.33 | 2.96 | 2.50 |
| | 50 | 4.74 | 4.63 | 4.50 | 4.31 | 4.09 | 3.79 | 3.47 | 3.05 |
| | 75 | 4.88 | 4.83 | 4.76 | 4.67 | 4.55 | 4.38 | 4.18 | 3.89 |
| | 100 | 4.93 | 4.90 | 4.86 | 4.81 | 4.74 | 4.63 | 4.50 | 4.31 |
| 7 m | 25 | 5.33 | 4.83 | 4.30 | 3.69 | 3.11 | 2.51 | 2.02 | 1.53 |
| | 32 | 5.88 | 5.49 | 5.06 | 4.53 | 3.97 | 3.34 | 2.80 | 2.20 |
| | 40 | 6.24 | 5.96 | 5.61 | 5.19 | 4.70 | 4.12 | 3.57 | 2.92 |
| | 50 | 6.49 | 6.30 | 6.05 | 5.72 | 5.33 | 4.83 | 4.33 | 3.69 |
| | 75 | 6.77 | 6.67 | 6.54 | 6.37 | 6.14 | 5.84 | 5.50 | 5.01 |
| | 100 | 6.87 | 6.81 | 6.74 | 6.63 | 6.49 | 6.30 | 6.07 | 5.72 |
| 10 m | 25 | 6.91 | 6.10 | 5.27 | 4.39 | 3.58 | 2.81 | 2.21 | 1.63 |
| | 32 | 7.85 | 7.19 | 6.47 | 5.61 | 4.78 | 3.90 | 3.18 | 2.42 |
| | 40 | 8.51 | 8.00 | 7.41 | 6.67 | 5.88 | 5.00 | 4.21 | 3.33 |
| | 50 | 8.99 | 8.62 | 8.17 | 7.58 | 6.90 | 6.10 | 5.32 | 4.39 |
| | 75 | 9.53 | 9.34 | 9.10 | 8.76 | 8.34 | 7.79 | 7.19 | 6.37 |
| | 100 | 9.73 | 9.62 | 9.47 | 9.30 | 8.99 | 8.62 | 8.20 | 7.58 |
| 15 m | 25 | 8.97 | 7.65 | 6.40 | 5.14 | 4.07 | 3.10 | 2.40 | 1.73 |
| | 32 | 10.64 | 9.46 | 8.24 | 6.91 | 5.68 | 4.49 | 3.55 | 2.64 |
| | 40 | 11.88 | 10.91 | 9.84 | 8.57 | 7.32 | 6.00 | 4.90 | 3.76 |
| | 50 | 12.84 | 12.01 | 11.23 | 10.14 | 8.97 | 7.65 | 6.46 | 5.14 |
| | 75 | 13.96 | 13.55 | 13.05 | 12.36 | 11.55 | 10.51 | 9.45 | 8.10 |
| | 100 | 14.40 | 14.15 | 13.84 | 13.39 | 12.84 | 12.10 | 11.28 | 10.14 |

# Setting Up Front Projection

Shadow and halo effects in front projection result from A, the physical diameter of the lenses; B, misalignment of camera and projector. Problems arising from A may be reduced but not eliminated (see previous page). The two factors combined have a cumulative effect and errors of one kind decrease the permissible tolerances of the other.

As a cameraman often requires to set foreground objects as close to the camera as possible, it is necessary to line up the equipment with a great deal of accuracy. As with 'the nearest object distance' (see table on previous page) the wider the lens angle and the greater the screen distance the greater the tolerances in camera/projector alignment.

## Lining-up

1. Set up the camera, projector and screen in their operating positions.
2. Set up targets made of patches of Scotchlite material on stands. Place them at the *f*16 minimum object distance (see page 112) on the centre line and the left and right hand extremes of the foreground objects.
3. Switch on the projector and twist the targets until they have the same reflectivity as the background screen when viewed through the camera.
4. Adjust the camera/projector alignment until no shadow is cast by any target.
5. If the camera is on a nodal head, pan and tilt it to the degree required by the shot to ensure that no shadows appear.
6. If a zoom lens is used on the camera, set up the shot at the longest focal length to be used and zoom to the widest angle as a double check. It may be found that only a limited amount of zooming is possible without sliding the camera back and forth on the head to keep the front entrance pupil position constant.
7. Set the camera and projector on their working apertures. The exit pupil of the projector lens should not be larger than that of the camera. Check by comparing their relative depths of field, (in tables). The pupils are equal when depths of field are the same. If a shadow appears at this stage, it may be caused by the light source being off-axis in the projector.
8. Check colour balance between the background and foreground (see page 122).

SETTING UP FRONT PROJECTION

**Setting up a camera and projector**
Set three targets of Scotchlite material in front of the screen and turn slightly until the intensity of the light reflected from them is equal to that from the background.
1. Projector set too close to the screen relative to the camera, shadows on the outer edges.
2. Camera set too close to the screen relative to the projector causing shadows on the inside edges. (Note that there is more tolerance in alignment if a single subject is set astride the centre line.) A. Camera; B. Projector; C. Semi-silvered mirror; D. Scotchlite targets; E. Scotchlite front projection screen.
3. Mirror not at 45°, causing asymmetric shadows.

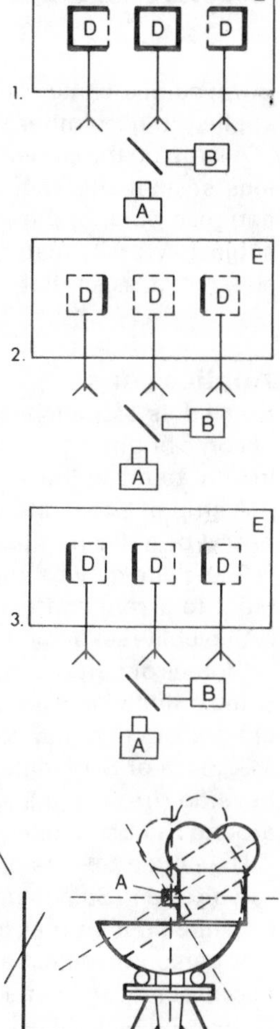


**Setting camera on a nodal head**
4, 5. Set camera on a suitable head with the optical axis of the lens level with the centre of rotation of the tilt and above the pivot point of the pan. Slide the camera back or forward until the two separated target objects set up in line with the camera do not move relative to each other when the camera is panned and tilted.
A. Nodal point of head and lens; B. Targets set up in line with the camera.

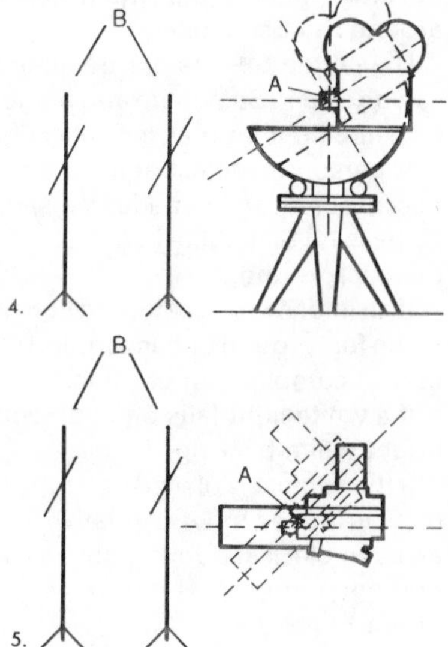

# Other Uses for Scotchlite

Because Scotchlite can reflect light directly back to the source, it may be applied to uses other than in process projection.

A lamp on the camera shines onto a partial mirror set at 45° in front of the lens so that the source of light beam, being coincident with the front entrance pupil of the lens, is reflected along the optical axis towards the subject. A totally matt black surface must be set on the opposite side of the sheet of glass so that unwanted light is not reflected back into the camera lens.

## Applications

Small dots of Scotchlite may be stuck to a studio night sky background to become bright stars. The intensity of the lamp on the camera may be varied in shot to make the stars 'twinkle'.

A strip of Scotchlite stretched between two points may become a 'laser beam' or a 'flash of lightning' etc.

Scotchlite may be set into windows of a studio set of a city or houses by night to give the effect of internal lights.

Scotchlite set behind coloured glass mosaics makes them shine brightly.

Clothes or part of a set or model may be made of Scotchlite and caused to shine brightly or change colour in-shot by projecting light sources of different and/or changing colours.

A piece of Scotchlite placed in or behind a glass of beer or other clear beverage makes the liquid sparkle. If necessary, the artist must drink around the Scotchlite.

Unwanted objects placed against a plain blue or white sky background can be clad in Scotchlite and made to disappear by shining a matching blue or white light so that the reflection merges into the background.

A person or object against a blue or white sky or other plain coloured background can be made to disappear, partially disappear or be cut in half by passing Scotchlite in front during a take to reflect a light which matches the background.

If an intense blue coloured light is used on the camera, areas of Scotchlite in the foreground or background can become 'blue backing' for subsequent optical combination printing.

If a white light falls on it, a Scotchlite backing can ensure a shadowless bright white backing.

If the subject is placed centrally about the optical axis of the lens, the light may be moved forward relative to the front entrance pupil of the lens so that an even black shadow is set around the subject. This is quite useful when photographing a white, or partially white subject against a light background.

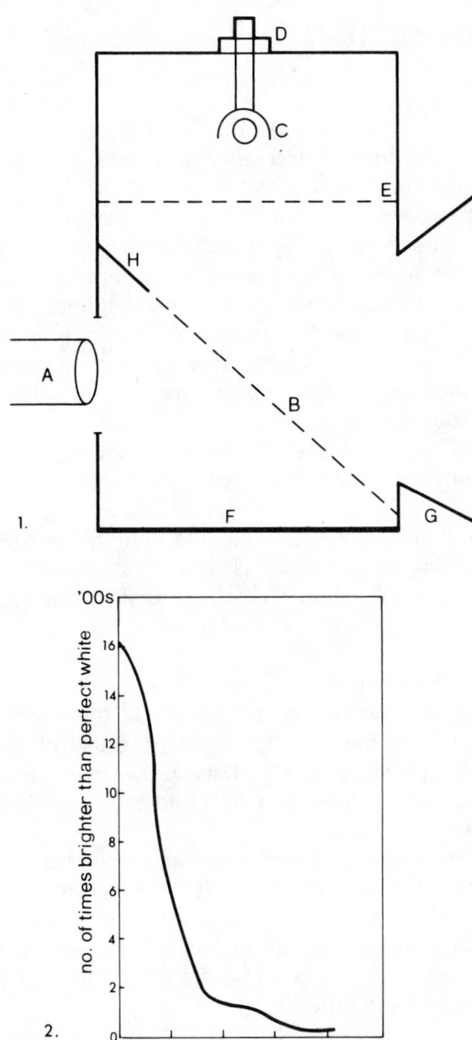


ADDITIONAL USES FOR FRONT PROJECTION MATERIAL

1. **Light box on camera front**
Schematic layout of light box: A. Camera lens; B. Partial mirror; C. Lamp and reflector; D. Adjustable bracket to position lamp precisely; E. Holder for lamp filter; F. Matt black surface; G. Sunshade; H. Shade to stop light from lamp falling on camera lens.

2. **Light intensity**
Graph showing how intensity of light reflected by Scotchlite falls off sharply unless it is coincident with the source. A $\frac{1}{2}$° difference in angle reduces the reflected light by 60 per cent.

*A long-standing form of process photography.*

# Back Projection

As its name suggests back projection involves projecting images from behind the subject onto a translucent screen placed between the foreground and the projector.

As with front projection, when using film plates, a special process projector must be used for back projection which incorporates a register pin movement, as accurate as that of a camera, and which will run interlocked with the camera, ensuring not only that both run at an identical speed but also that the two films remain stationary and are advanced simultaneously. To achieve the necessary accuracy of registration, prints for process projection (front and back) are usually made on negative-perforated filmstock.

Screens as large as those used for front projection cannot be used for back projection because of the light losses inherent in using a translucent screen.

Although the line-up between camera and projector in the back projection process is not as critical as with front projection, the projector lens must be reasonably square-on the screen if the screen is not to appear darker on one side than the other.

## Light and colour balance

An inherent problem with back projection is the effect of a central 'hot-spot' of light. This may be reduced by the use of very long focus lenses. Some studios have special back projection tunnels permitting 150–200mm focal length lenses to be fitted to the projector and 50 or 75mm lenses on the camera.

As with front projection, achieving correct lighting and colour balance between foreground and background is an inherent problem. This is usually tackled by a specialist technician. As a rough guide, the foreground lighting should be half that which appears to be necessary. This is to compensate for the fact that the projected background is only being illuminated half the time.

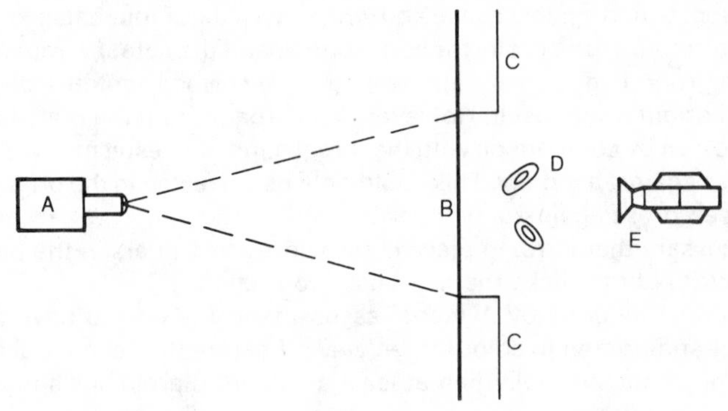


BACK PROJECTION

**A back projection set-up schematic**
A. Back projector with long focus lens; 200–300mm is quite normal; B. Translucent back projection screen; C. Solid parts of set; D. Foreground subject; E. Camera on dolly.

# Process Projector Colour Balance

An inherent problem with both front and back projection is that of colour balance. The colour balance of the plate must be controlled to match that of the foreground at the time of shooting.

This situation does not occur in the case of travelling mattes as any colour correction between foreground and background may be carried out subsequently, on the optical printer.

## Foreground and background colour balance

A typical process projection shot might be of, say, a woman in a white dress to be photographed against a blue and white sky. If the colour balance of the 'plate', the colour of the projector light source, and the colour temperature of the foreground lighting are all compatible, a correctly colour-balanced combined original will result. However, if the image from the plate has an overall red cast by comparison with the foreground, the result may well be a white dress against a red sky. This could only be corrected in the printing to give a green dress against a white sky.

It is necessary therefore, to place colour-correcting filters in the path of the projector light to make the necessary correction.

As this must be done by personal assessment it is well to have a test scene shot and printed in colour to be viewed before the scene is actually shot. If time is not available then at least a colour Polaroid still should be shot.

If the foreground and background parts of the scene both contain areas of pure white it may be possible to assess the colour balance accurately by the use of a spot colour temperature meter.

Any colour bias on the plate must be judged to be towards one of three primary colours, red, green, blue, or three complementaries, cyan (green-blue), magenta (red-blue) or yellow. A filter of the opposite colour and of suitable density must be used to correct.

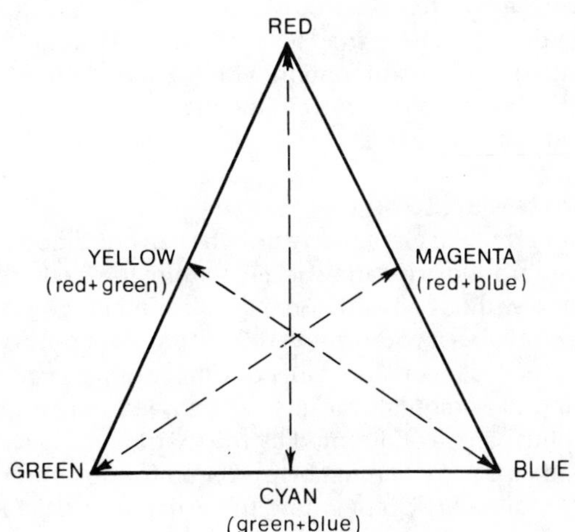


LIGHT COLOUR TRIANGLE

To lighten a colour, use a filter of similar colour. To darken a colour, use a filter of complementary (opposite) colour.

# Travelling Matte or Blue Backing

Many producers prefer to use travelling matte in preference to other process systems because it requires a minimum of fuss on the production floor.

Foreground action is played in front of a plain, brightly lit, blue background screen. Any similar blue colouring or reflections in foreground scenery, props or clothing should be avoided. Yellow filters are sometimes placed over the foreground lighting, the yellow cast being removed subsequently in the printing. The simplest method of making a combined negative is to make male and female high contrast B & W mattes from the foreground negative in which the foreground and background areas will be dense black and clear white.

The composite print, or CRI, is made by passing the foreground negative through a bipack printer in combination with the female matte and repassing it through the printer to expose it to the plate negative in combination with the foreground male matte.

## Travelling matte variations

By the use of a separate auxiliary matte drawn to the outline of a part of the background, a foreground artist may be set 'within' the background image. By this means, and without an additional duplicating stage, artists may be made to pass behind objects or through doors that exist only on the plate.

By the same means, suspension wires or other supports carrying actors or objects in flying or 'weightlessness' scenes, need not be hidden at the time of shooting but can be obliterated by the matting process. In this case the handdrawn auxiliary matte must be prepared frame by frame but need only be accurate at the very point where the artist, or article and support, touch.

Two or more foreground scenes shot against blue backings may be combined with one background. In this manner one artist may play two or more parts and walk around himself without the need to use a double shot 'over the shoulder'. By the same means 'giants' and 'midgets' may play a scene together or a person be made to grow or shrink.

## Blue backing by front projection

Part of a person or object may be made to disappear by shooting against a blue backing and covering the unwanted portion with Scotchlite front projection material. Onto this, a blue light is shone via a semi-silvered mirror as in front projection. The Scotchlite material may even be drawn across in-shot to give a disappearing effect.

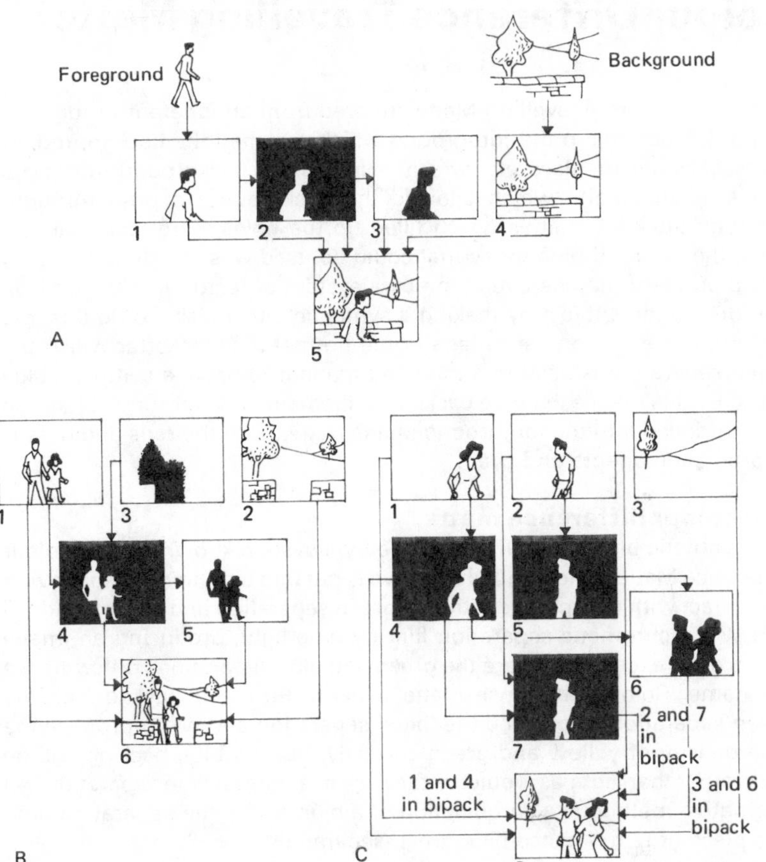


A. SINGLE FOREGROUND IMAGE SUPERIMPOSED IN FRONT OF A BACKGROUND IMAGE

The foreground subject is photographed against a blue backing and the background photographed in the normal manner.
1. Camera negative of foreground; 2. Camera negative of background scene; 3. B & W female matte of foreground subject; 4. B & W matte; 5. Combined CRI or positive made by printing 1 in bipack with 3 and 2 in bipack with 4.

B. FOREGROUND IMAGE SET BEHIND A SECTION OF BACKGROUND IMAGE

1. Camera negative of foreground; 2. Camera negative of background scene; 3. Hand-drawn matte around section of background scene; 4. Female matte made from foreground negative and auxiliary matte; 5. Combination print made by printing 1 in bipack with 4 and 2 in bipack with 5.

C. ONE FOREGROUND IMAGE PASSING IN FRONT OF ANOTHER FOREGROUND IMAGE AGAINST A SEPARATE BACKGROUND IMAGE

1. Camera negative of dominant foreground subject shot against blue backing; 2. Camera negative of secondary foreground subject shot against blue backing; 3. Camera negative of background scene; 4. Female matte of 1; 5. Female matte of 2; 6. Male matte of 4 and 5; 7. Female matte of 2 and 4; 8. Combination print of 1 printed in bipack with 4, 2 printed in bipack with 7, and 3 printed in bipack with 6.

# Colour Difference Travelling Matte

In the early days, Travelling Matte suffered from an inherent problem in that any blue area in the foreground which matched the background, or areas of water or glass etc. which reflected the background, produced 'holes' in the matte which allowed the background to print through; although auxiliary mattes which filled up the holes were (and still are) used, the range of blue tones that could be used was strictly limited.

The problem may be overcome by a system of restoring blue tones in the foreground subject by making a synthetic blue mask. To do this, red and green separations are made in the normal manner, after which the green separation is used in a modified manner to make a synthetic blue mask. By this means the blue backing will remain dark but some 'blue' can be introduced into the foreground image to mix with the reds and greens to produce magenta and cyan.

## The colour difference matte

The synthetic blue separation is made by a system known as the 'Colour Difference Matte'. The original negative is put into a printer in bi-pack with (in contact with or together with) the green separation and a print made in B & W panchromatic separation film by blue light, producing an image that is clear except for where the green and blue separations differ; hence the name 'Colour Difference Matte'. This in turn is bi-packed with the green separation to become the 'blue' separation and will reproduce the blue density of yellow and green correctly, but the blue backing will be dark rather than light as would be the case in a normal blue separation. An alternative method is a combination of a blue and green separation onto one piece of film – hence blue-green separation.

To produce the mattes, a special Hi-contrast blue separation matte is first made from the original. This will have a clear background and a dense foreground. This matte is then printed in bi-pack with the original negative using red light and panchromatic film to produce a female matte from which a male matte can then be made.

## Making the composite negative

To produce the final composite negative the three separations of the foreground image are bi-packed in turn with the female matte and printed onto internegative stock, followed by the background separations which are exposed in contact with the male matte.

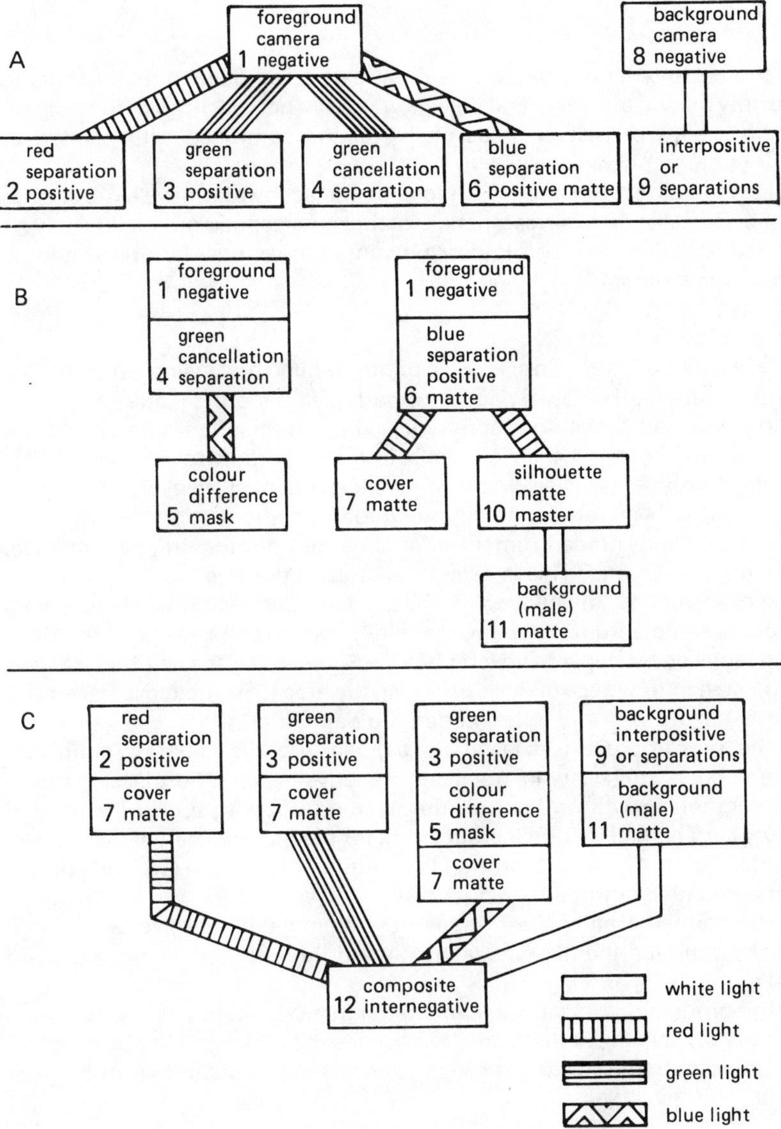


Flow chart and system as operated by Roy Field and R. A. Dimbleby

A. **First stage:** manufacturing component from foreground and background camera negatives.
B. **Second stage:** manufacturing secondary components from foreground camera negative and first stage components.
C. **Third stage:** combining components to make composite internegative.

# Glass Mattes

Many creative films depend to a great extent, on painted matte shots to portray places, buildings and objects which are not practicable to construct. Such scenes consist of areas of live action combined with painted or still-photographed scenes.

Low buildings with busy street foreground action may have extra storeys put on to make them skyscrapers. Distant snow-capped mountains may be added to a desert foreground or unwanted areas may be obliterated by being overlaid etc.

## In practice

The live part or parts of a scene are photographed first, using a register-pin camera mounted on a very solid support, (see Shooting Plates, page 130) and processed in the normal manner. These parts of a scene should not contain movement which overlaps into the portion to be matted. So, exclude people, vehicles, water or even smoke, dust, rain, clouds or heat haze, etc. which would not continue into the matte section.

A well colour-graded (timed) print is made from the original negative, using a step printer. This is called the 'master positive'.

A clip from the 'master positive' is put into a projector, which may be a process camera from which the rear film pressure plate has been removed and replaced by a prism and lamphouse (Rotoscope). The selected frame is projected onto a piece of glass on which the area to be matted is traced and the areas of the scene to be matted are painted on.

The 'master-positive' is laced into a bipack process camera in emulsion-to-emulsion contact with duplicate negative stock. The glass matte is photographed with no light on the painting but with a well illuminated white card behind to form a printer light for the required part of the original scene. These are passed through the camera, the film is rewound and the dupe negative re-threaded on its own.

The painted area of the glass is now illuminated, the live action area blacked out and the dupe negative passed through the camera a second time.

The production of believable painted mattes is a highly skilled job done by very few people in the world. But scenes to be matted may be supplied by any cameraman with a register pin camera and a sturdy tripod.

GLASS MATTES

1. Original scene is photographed with a register pin camera set on a rock steady support. The negative is developed and a print made on a step printer.

2. Part of scene to be matted in is painted onto glass after frame of original has been projected on to trace a suitable dividing line.

3. The master positive (B) is laced up into a bipack camera (C) in emulsion to emulsion contact with unexposed duplicating stock (A). The clear area of the glass matte (D) permits light to pass through from an illuminated white background (E) which acts as a printer light.

4. The partly-exposed duplicating stock (A) is re-laced into the camera (C), which photographs the painted section of the glass matte (D), now illuminated from the front, while the background (E), is made dead black.

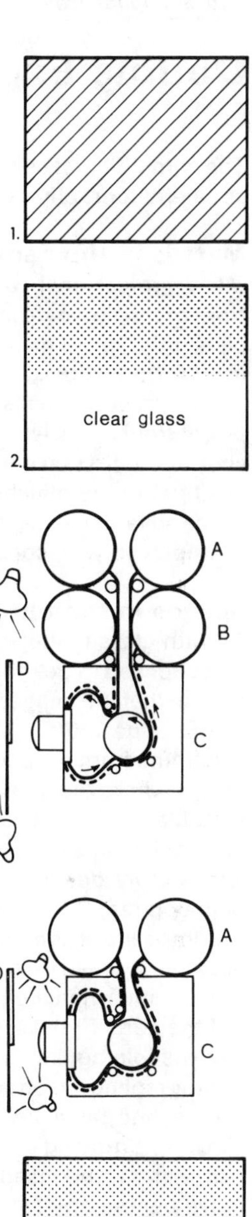


5. Result: combined scene, partly original, partly painted.

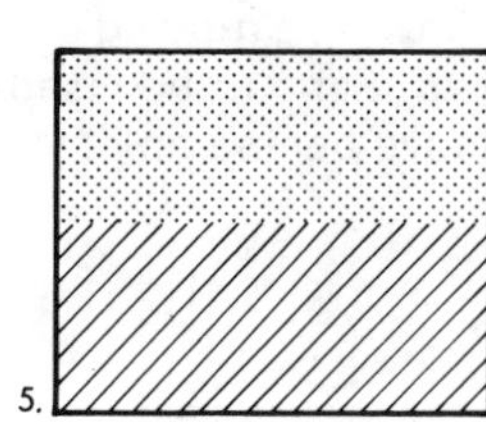

# Double Exposure Mattes

The combining of several live action scenes into one is an art almost as old as cinematography itself. There are many ways of achieving the same end.

## Mattes in the camera

Many professional cameras are equipped with slots immediately in the front of the film gate into which mattes may be slid. Being close to the film plate, this type of masking has a sharply defined edge. A matching matte is used for the subsequent exposure.

If a soft, out of focus transition is required, a matte is placed in the matte box in front of the lens. When this technique is used, sufficient film must be shot of the first take to allow for some trial-and-error tests to be made in positioning the matching matte for the second exposure.

The advantage of matting in the camera is that only one generation of film is employed with consequently no loss of image quality.

## Mattes in the optical printer

As with glass mattes (page 126) the live action elements of the scene are first exposed in the normal manner, positioning the camera on each occasion so that the images are correctly placed in the frame area (see Shooting Plates, page 130).

A frame from the background scene is projected on to a white card, a line drawn where the images are to be joined and one side of the line painted matt black. This is then photographed, either in focus, if a clear-cut join is required (to line up with the edge of a building, etc.) or out-of-focus if a soft join is more desirable. From this negative matching mattes are made on high contrast B & W stock.

Composite second generation negatives may be made from the original negative (printed in contact with the related matte), onto a CRI internegative or from intermediate positives of the original negatives.

If the parts of the scene to be matted-in contain movements which butt into the joining line, a whole series of mattes may have to be drawn and photographed on an animation table.

A second (auxiliary) matte may be made from the background scene allowing a moving object which may have been photographed by a travelling matte to be inserted 'into' a scene. Thus, a person could be seen to move behind an object which originally formed part of the background.

CAMERA AND PRINTER MATTE SHOTS

**Mattes in the camera**

1. Original scene is photographed with unwanted part of scene marked off.
2. Original negative is re-passed through the camera with the complementary section marked off.
3. Result: combined image on original negative.

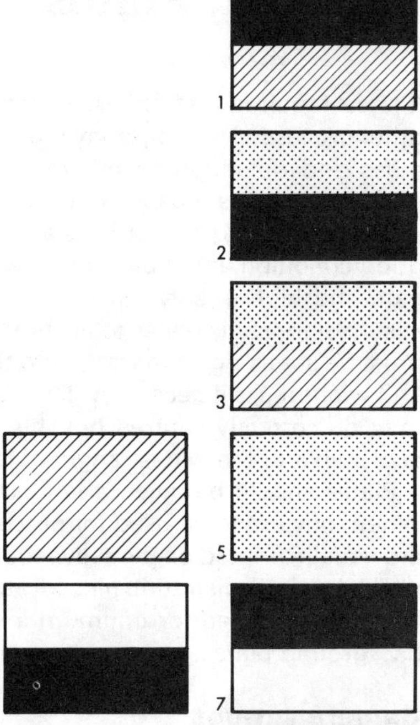


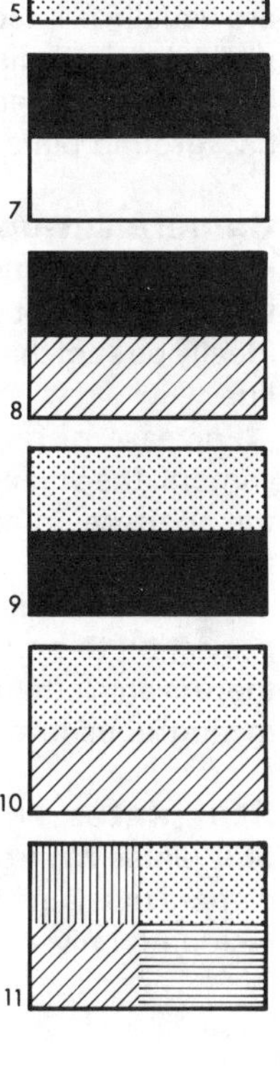


**Mattes in the optical printer**

4, 5. Original scenes are photographed, stock shots may be used.

6, 7. A suitable black and white matte is drawn, photographed on high contrast stock and a complementary print made.

8. A duplicate is made from one original scene in contact with a black and white matte.

9. The second scene, in contact with the complementary matte, is printed onto the unexposed portion of the duplicate.

10. Result: combined image on CRI or duplicate.

11. A number of complementary mattes may be made for a multi-image combination.

# Shooting Plates

'Plates' required for front projection, back projection, or travelling matte or other processed photography must have the highest possible image steadiness and definition to minimise the inevitable degradation when re-photographed as a background.

For static background plates a register-pin 35mm camera, in particularly good condition, must be used, preferably fitted with a full (silent) camera gate and high quality lenses. A heavy duty head on a robust tripod is essential for static plates to ensure that there is no movement. The camera must be set as rigidly as if it were the ground itself, using chains, weights and windbreaks if necessary. If the camera has a full aperture gate the lens may be correctly centred but this is not necessary if some limitation is acceptable in the widest angle of lens to be employed.

Plates should be given a full exposure so that they do not lack detail. When they are to be used for process projection or bi-pack mattes, plates are usually printed on 'negative' perforated film stock, using a step printer which has its registration pins located as they were on the original camera. There is less need for optimum image steadiness in the case of moving background plates.

## Camera set-ups

Process shots must be carefully planned to ensure that plates match when the action is shot at a later date. Where a plate includes a horizon it should ideally be shot three times with the horizons above, below and on the centre line.

For ease of recall each plate should be slated with the height of the camera, the angle of tilt, the direction of the light, the lens focal length, and the aperture, in addition to the usual information. An inclinometer and a large protractor may be used for measuring angles.

Plates, particularly those with a full aperture, may be flipped over in projection to provide a reverse direction or angle. This cannot be done if there is any lettering or recognisable object in the picture which would look incongruous if reversed.

## Still plates

Still plates for front projection are usually made with a 10 × 8in format camera to ensure that the grain of the plate is not obtrusive when close-ups including only a small section of the background are shot.

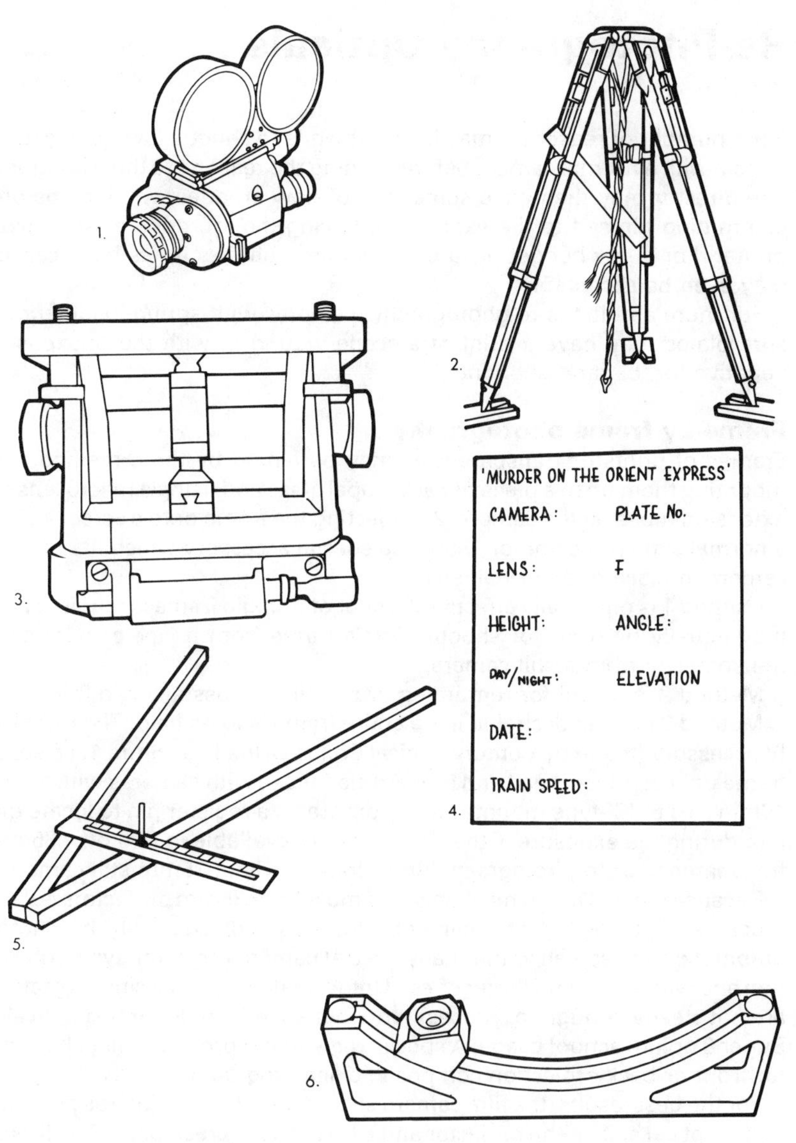


EQUIPMENT FOR SHOOTING PLATES

1. Register-pin camera; 2. Rugged tripod; 3. Rugged head with locks; 4. Detailed marker board; 5. Protractor; 6. Inclinometer.

# Re-Photography Opticals

For a number of reasons it may be worthwhile to shoot a low-quality, trial optical with a normal camera before sending the real one to the laboratory. The director may then have some idea of how an effect will look before committing himself to the expense of having the job done in an optical printer. For a low budget picture, 're-photography' as it has been called, may even be acceptable.

For more ambitious re-photography, cameras with spring-loaded pressure plates may have a print of a scene wound in with the unexposed negative for 'bi-pack' shooting.

## Frame by frame photography

Frames of prints or transparencies may be filmed by three methods: 1. mounting them on to a piece of backlit opal glass and using a macro lens or extension tubes on the camera, 2. projecting the frame onto a screen using a normal slide projector, or 3. using a special accessory which fits on to a camera in place of a normal lens.

Method 1 is only really effective if a shot of a single frame is required and may equally be used for shooting off a frame from a cine-camera or a transparency from a still camera.

Method 2 is useful for re-framing or panning across a single frame.

Method 3 is most desirable if the entire frame is to be held. The Duplikin IV accessory (made by Century Optical Co.) is virtually a 16mm 1:1 freeze-frame optical printer head and is available for use with cameras with either 16mm Arri or 'C'-type mounts and incorporates a register pin to locate the film during the exposure. Other Duplikins are available to film off a 35mm transparency or to photograph 16mm frames with a 35mm still camera.

Because the pull down mechanism of most 16 or 35mm projectors is very much faster than that of a camera, film thus projected may be photographed off the screen without any special camera/projector synchronisation and with little or no flicker effect. Unfortunately, normal cine projectors are not steady enough for process work but are adequate for test opticals, Super 8 or film school usage. A special register pin projector must be used for front or back projection composite cinematography.

For the best results, the film camera should be set up as close as possible to the optical axis of the projector and a high-grain screen used. The flicker blade from the projector may be removed to give more light. The camera speed may be varied to give speeded-up or slowed-down motion.

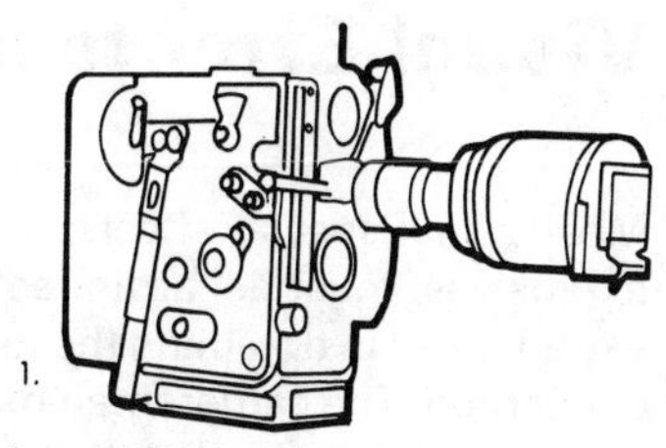

1. Duplikin II accessory for filming from 35mm slides.

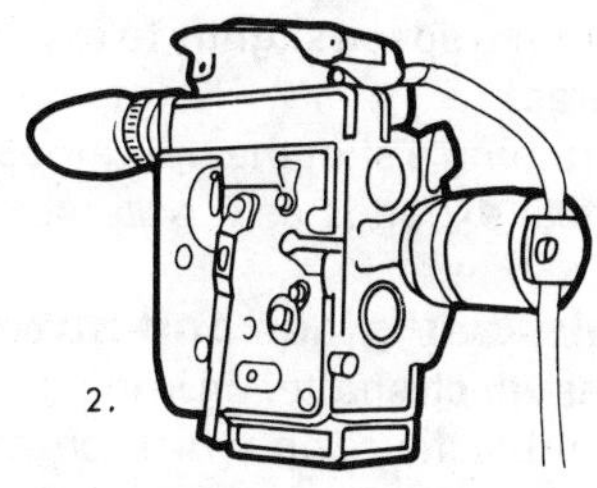

2. Duplikin IV accessory for filming from a 16mm frame.

3. Cine camera set up alongside a cine projector.

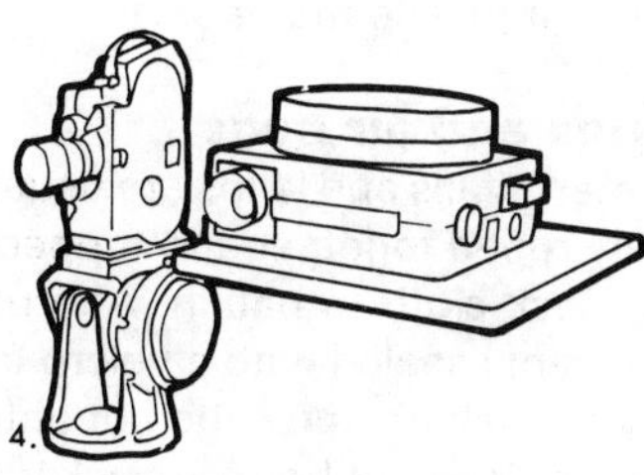

4. Cine camera set up alongside a slide projector.

# Visual Effects and Stunts

Most visual special effects, and stunts, especially those involving fire or explosives, must be 'arranged' by specialist technicians who are not only experienced in obtaining the required effect, but can ensure the safety of all concerned. The cameraman must work closely with the special effects and stunt men to ensure that their work is exploited to its maximum and that the means of its achievement are not revealed. It may be advantageous to vary the camera speed slightly to expand or contract time and to use two or more cameras.

In framing a shot the operator should allow for the fact that in most cases an effect expands as it occurs.

## Breakaway glass and pottery

Bottles which shatter on impact without risk of personal injury are made of a special plastic resin which crumbles rather than splinters when broken. They are available in a wide variety of forms and may be manufactured to match existing props. Sheets of 'glass' for windows may be made from a similar material but it is not ideal especially for medium and close shots, as it does not break and splinter in a similar manner to glass. When real glass is used a very thin picture framing type is employed and is pre-scored or even cut with a diamond cutter beforehand.

## Breakaway furniture

Where possible, props required to break in fight sequences should be made of balsa wood and weakened where the breaks are intended. Real furniture may be completely cut through as it is intended to come apart and lightly held together with very small dabs of glue or toothpick 'dowels', taking care to camouflage the join before impact.

Props which have to be thrown about may be made of vacuum-formed lightweight plastic material.

## Box rigs and air bags

Stunt men's falls and leaps are usually on to piles of large cardboard boxes securely roped together and topped with a mattress or other soft material and a black cloth tarpaulin. Alternatively, an airbag which incorporates release vents sealed with a velcro tape is used. Either system absorbs the energy of a falling body and not only cushions the fall but ensures that the stunt man does not bounce back into shot. Cardboard boxes and air pads may be 6ft high and more than 20ft in width, for which space must be allowed when the sets are designed.

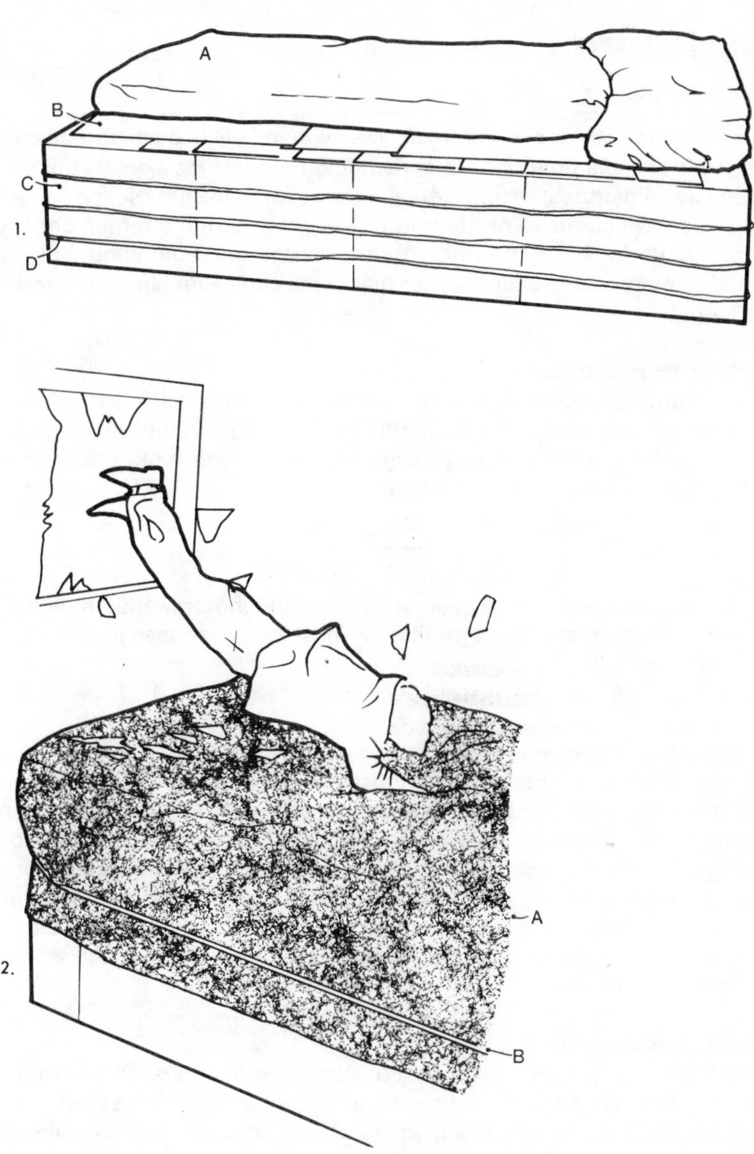


STUNTS WITH SAFETY AND EFFECT

1. **Typical cardboard box stunt fall rig**
A. Soft mattresses; B. Folded cardboard; C. Layers of empty cardboard boxes; D. Ropes holding boxes together; G. Black cloth to ensure no reflection in flying glass; F. Rope holding black cloth.
2. **Stunt man diving through glass window**

# Miniatures

Dams bursting, ships sinking, trains crashing, oil wells gushing and cities destroyed are examples where scaled down models may be used advantageously. Miniatures, as they are called, may form an entire scene containing objects which move or are part of a scene where the remainder is of normal scale, or be matte printed into a scene and combined with live action. Miniatures are best used for quick cuts and never left on the screen too long.

## Mobile miniatures

Because gravity cannot be miniaturised, models which fall, crash, blow up etc. must be slowed down in the filming by speeding up the camera by the 'square root of the scale' in order to make the movement look natural on the screen. The formula may be written:

$$\frac{\text{object size}}{24\times \sqrt{\text{size of model}}}$$

Thus, a 1/16 scale ($\frac{3}{4}$in = 1ft) miniature must be photographed at 96fps.

The speed at which mobile miniatures must be moved may be calculated using the formula:

model speed (ft per sec) = real speed (mph) × scale$^2$ × 1.47

or,

model speed (m per sec) = real speed (kph) × scale$^2$ × 0.28

When an object is seen to fall or crash it is important that its weight and strength should also be to scale. In such a case it is usual to build to one quarter natural size and pre-weaken the structure, to make the shot look realistic. For shots involving water (drops being indivisible) 1/16 is considered to be the smallest practical scale even if wetting agents, detergents, electric fans and wave generating paddles are used to create realism.

Models must be photographed from the same viewpoint as if full size and not looked down upon.

## Static miniatures

Miniature models of the upper part of a building or of a ceiling, complete with miniature chandeliers, may be hung in front of the camera to marry up with full-size sets which do not have an upper section. They have the advantage over process mattes in that they may be panned off or onto, using a camera mounted on a nodal head.

Miniatures of buildings may be placed just beyond windows or other openings of a set to give a false idea of distant perspective with further objects being progressively scaled down even more, to help overcome the depth of field problem. The effect of distance may be increased by the introduction of 'haze', reducing model detail and 'greying down' the furthermost objects.

MINIATURES USED IN FILMS

1. 30ft (9m) 1/12 scale model of the 1906 warship *Konigsberg* used in the film, *Shout at the Devil.*
2. 44, 22, 11 and 5½in (1120, 560, 280 & 140mm) 1/24, 1/48, 1/96 and 1/192 scale models of the *Eagle* spaceship used in the TV series, *Space 1999*.

*Shooting a TV set without showing a bar line.*

# Shooting a TV Screen

There are occasions when it is necessary to shoot a scene which includes a practical TV set in shot. In a feature film, the TV picture is invariably from a VTR source. But in documentary or news work, the TV may be receiving 'off-air'.

## Bar problem

If a TV set is shot with a normal reflex camera a 'hum', 'frame', or 'shutter' bar is seen in the viewfinder and, subsequently, on the film. This is a broad dark or light band which may move rapidly and successively up or down the frame, or remain stationary. It is caused by the film photographing more (if the band is light) or less (if the band is dark) than a complete electronic scan of the TV screen. If the camera runs more slowly than the TV screen the band moves downwards, if faster, upwards. If it remains static then the film camera is running and scanning at the same speed. Cameras (eg: Arri 35 IIC) which have a black 'anti-flicker' segment on their rotating mirror shutters show an additional broad black band in the viewfinder.

By photographing a TV set correctly, with an ordinary film camera, bar lines which show up on the film may be totally eliminated. But the image quality is somewhat coarse, as only alternate scans are filmed. Such quality is adequate for shooting a practical TV but not for serious kinescoping.

## TV time base

The scanning rates of broadcast TV, or from VTR machines, are controlled by 50 or 60Hz quartz crystal oscillators, as appropriate. To remain in constant syncronism the camera used to film TV pictures from these sources must be electronically interlocked with the TV time base signal or be quartz crystal-controlled.

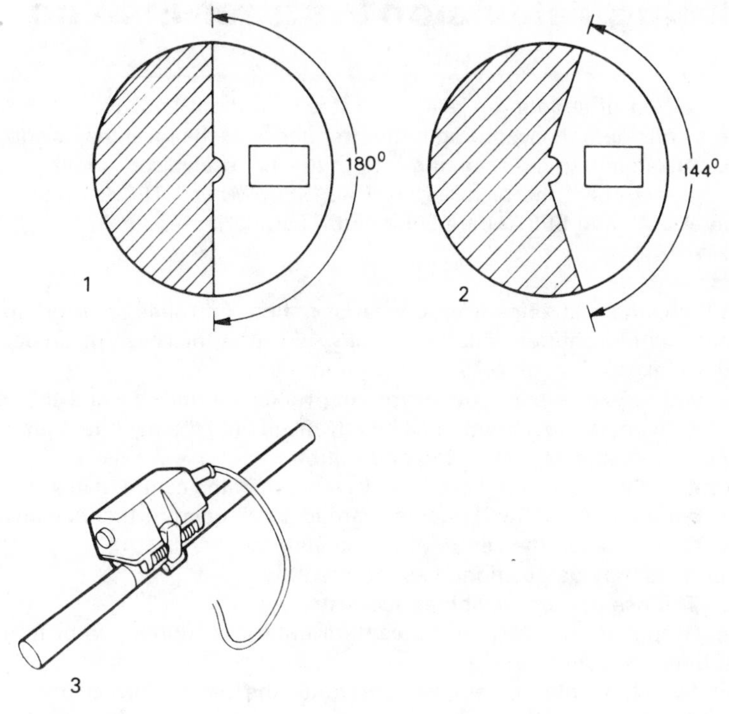

SHOOTING A TV SCREEN

1. Shutter suitable for: 50Hz 25fps, 60Hz 30fps; 2. Shutter suitable for: 60Hz 24fps; 3. Phase advancer to phase crystal controlled motor with TV scanning; 4. 50Hz TV scans, 25fps camera with 180° shutter, black areas show shutter closed; 5. 60Hz TV scans, 24fps camera with 144° shutter.

*Bar-lines are not really necessary.*

# Filming Television Pictures in Sync

The method of shooting a television screen without showing a bar line differs considerably according to the frequency of the local mains supply and, consequently, the scanning rate of the TV receiver or monitor.

It matters not if the pictures are PAL, SECAM, or NTSC, although, in general, PAL and SECAM are 50Hz and NTSC 60Hz.

## 50Hz

Run the camera at 25fps from crystal or mains control as appropriate. A camera with a shutter which is *exactly* 180° (the thickness of a coat of laquer makes a difference) shows no hum bar.

A camera with a plain mirror rotating reflex shutter shows a thin bar line. The camera movement must be advanced until the bar line is just out of view or visible equally at top and bottom.

A camera with an 'anti-flicker' black segment in the centre of its rotating mirror shutter (Arri 35 IIC) shows a broad black band in the viewfinder. This must be set in the centre of the screen.

Bar lines may be positioned as required by:

1. The use of a phase shifter accessory.
2. Stopping and starting the camera until it is running with the bar positioned correctly.
3. Switching the camera to 24fps until the bar is correct and then switching over to 25fps.

## 60Hz

1. Run the camera at 24fps (crystal or mains control as appropriate). A camera with a shutter which is *exactly* 144° films four TV scans out of a possible ten with the bar line eliminated or minimised. 2. Run the camera at *exactly* 30fps (crystal or mains as appropriate) and observe the same rules as for 50Hz/25fps. The film may be slip-frame printed missing out every fifth frame to correct the frame rate exactly 24fps and to be suitable for synchronisation with sound recorded at that speed. Provided that the scene shows no very fast moving object, the picture will not be jerky (all the original Todd-AO 70mm films (*Oklahoma, Around the World in 80 Days,* etc) were shot at 30fps and subsequently step-printed at 24fps when being optically reduced for 35mm anamorphic presentation).

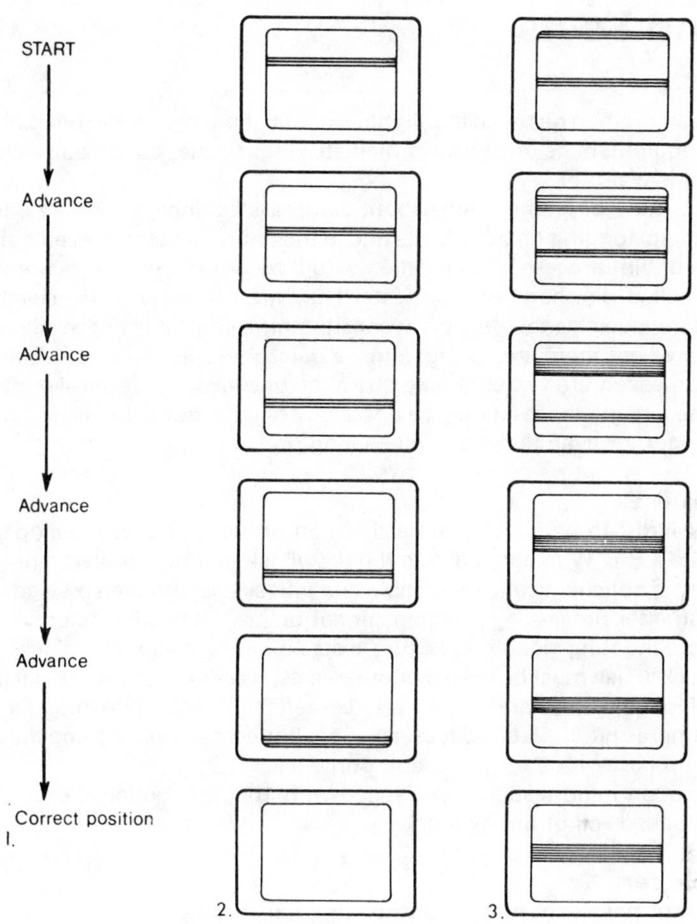


FILMING A LIVE TV (Kinescoping)

1. Phase shift action; 2, 3. Frame bars and anti-flicker segments as seen in camera viewfinder; 2. Camera with mirror shutter which has no anti-flicker segment; 3. Camera with mirror shutter which has an anti-flicker segment.

# Hand Tests

If the cameraman has even the slightest doubt about the proper functioning of his equipment he must take immediate steps to check and double check that all is well.

There may be some query about camera steadiness, focus, framing, scratching, fogging or other faults and in these circumstances even a short length of film processed on location would be useful. Such tests are variously called dip, hand or slop tests. They are not suitable for checking exposure levels as the necessary sensitometric control is not available.

Even where there are no problems, a quickly developed piece of film is sometimes required when setting up a shot for subsequent optical printing. In these circumstances it may be necessary to cut a frame to place it in the viewfinder, or even the gate, while lining up.

## Use B & W

Colour film can be developed easily in an ordinary B & W developer to produce a B & W image, but has the drawback that the jet black antihalo backing is not chemically removable and unless it is removed by vigorous rubbing after processing, the film cannot be projected satisfactorily.

When shooting double exposure tests to check image steadiness the filmstock in use must be used. For other tests, B & W stock is quite satisfactory. If longer pieces are not to hand, cassettes of 35mm film intended for still cameras purchased in a local shop are better than nothing and may be made up into a loop for projection purposes.

If location hand tests are envisaged with 16mm shooting take a 100ft spool-loaded roll of B & W stock.

## Developer

There is no need to be fussy about quality when making a dip test and a high contrast fast acting developer, as used for printing paper, is likely to be as good for the purpose as a fine grain negative mix. A fixing solution should be used that stops development, fixes the image, clears the film rapidly and hardens the emulsion, so that it may be washed in cold water.

Washing should be reasonably thorough so that all fixer is removed, with no residue remaining which might contaminate equipment and other filmstock. This is particularly important if using the film for a cut frame or matt.

Most developing tanks take only 5ft (1.6m) lengths of film but 100ft (30m) tanks are available. In an emergency, the chemical solutions can be put in a hotel bath or wash basin and the film processed by hand. (Don't forget to wash the bath or basin out well after such use!)

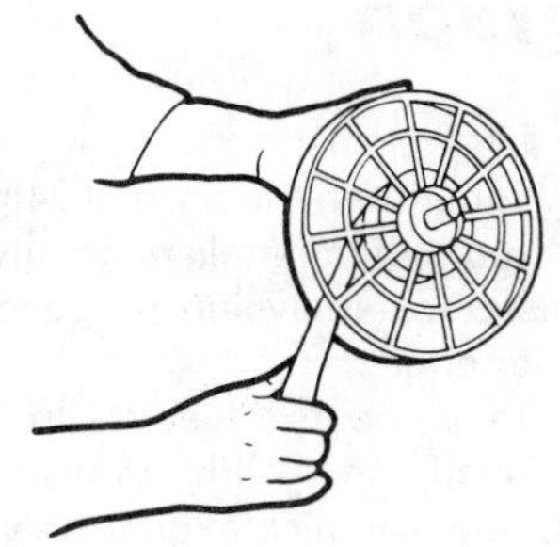
1.

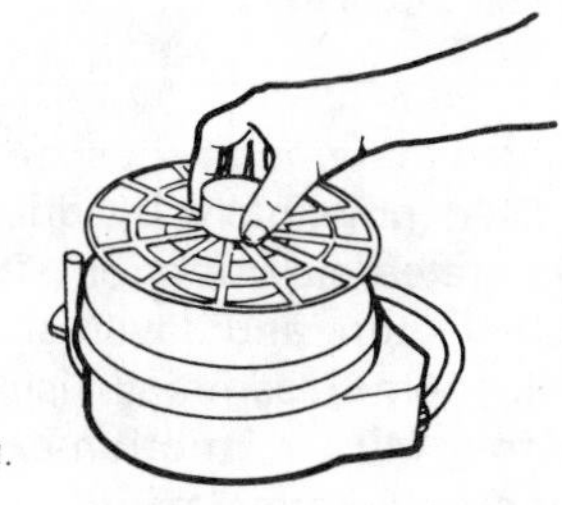
2.

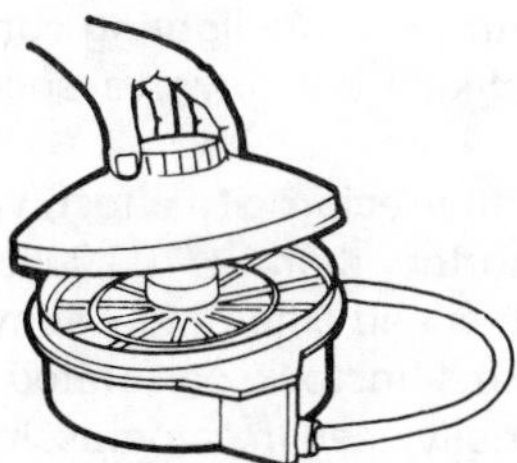
3.

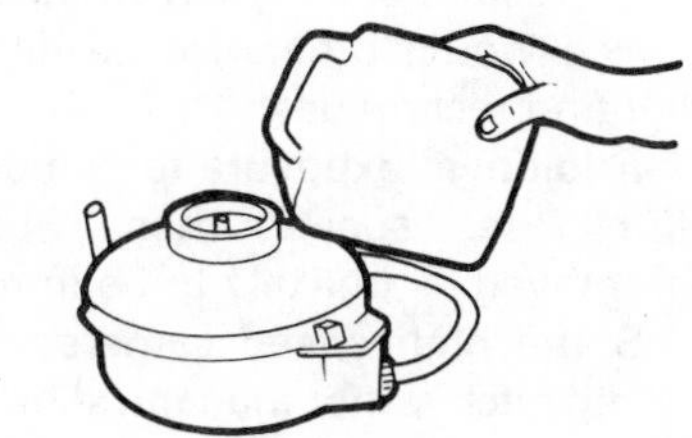
4.

LOCATION HAND TESTS

**Preparation**
1. Film is wound onto spiral holder (in complete darkness); 2. Spiral is placed in developing tank (in complete darkness); 3. Light tight lid is placed in position (in complete darkness); 4. Pre-prepared developing solution is poured into tank.

**Developing procedure**
A. Development, with agitation $1\frac{1}{2}$–12 minutes depending upon type of developer; B. Intermediate wash with water of roughly the same temperature as the developer; C. Fix in an acid hardening solution, usually about 5 minutes; D. Wash in running water for 5–10 minutes; E. If necessary, buff off jet anti-halo backing; F. Dry.

# High Speed Cinematography

Cameras that operate at speeds many times higher than the normal 24fps are used principally by documentary and sports film makers to give exaggerated slow motion effects, by scientists for analysing purposes, and by feature and advertising film makers for effect.

With 16mm cameras using intermittent movements, speeds up to 1000fps (40 × motion-expansion) are available. With optically-compensated movements, up to 10,000fps. Cameras which expose only a half- or a quarter-high frame may be operated up to 20 or 40,000fps, respectively.

Cameras using 35mm film with an intermittent movement are available for operation at 120fps (5 times normal speed) and 360fps (15 times), and over 2000fps with an optically compensated movement.

## Film for high speed

High speed colour filmstocks are pre-eminently suitable for high speed cinematography. Sufficient exposure is always a problem and films which are nominally rated at EI 250 or EI 400, but which may be force-processed to EI 1600, make certain shots possible with tolerable lighting conditions which would otherwise be difficult to achieve and involve specialised lighting techniques.

Additional exposure to compensate for the reciprocity effect, whereby film loses speed when exposures shorter than 1/100,000sec are employed, is unlikely to be in normal high speed cinematography.

Some high speed cameras may require filmstock perforated with a 'long' pitch rather than the short pitch normally used in cameras. It is most important to check that the perforation is correct as a film jam at high speed may severely damage a camera both mechanically and electronically.

## Lighting

In order to provide sufficient light for the very short exposure times inherent in high speed cinematography, wide aperture lenses or high lighting levels must be employed. In scientific cinematography, and where the action is contained within a small area, the high intensity lighting can be localised. A simple way to concentrate light over a very small area and at the same time eliminate heat is to shine a spotlight through a flask of clear water. The water acts as a condenser and a heat absorber.

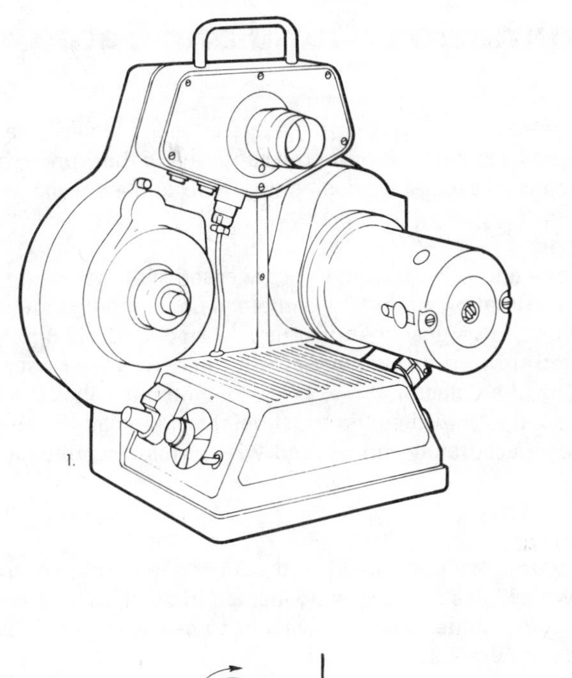


HIGH SPEED CINEMATOGRAPHY

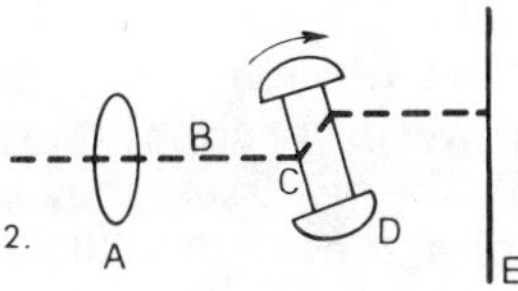

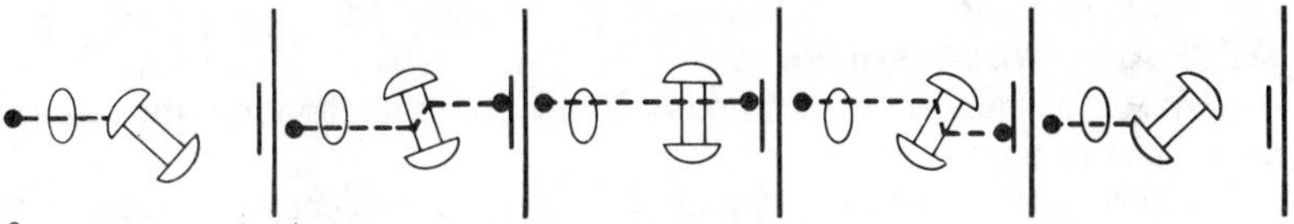

1. Hycam optically compensated high speed 16mm camera. Capable of 11,000 fps full frame 16mm, 22,000 fps $\frac{1}{2}$ frame and 44,000 fps $\frac{1}{4}$ frame.
2: A. Camera lens; B. Light path; C. Rotating optically flat glass plate; D. Drum shutter; E. Continuously moving film.

**The principle of optical compensation**
3. Light path marked by shutter; 4. Light path refracted upwards by glass plate; 5. Light passes directly through glass in perpendicular position; 6. Light path refracted downwards; 7. Light masked off by shutter.

# Animation Stand and Table

Animation drawings, diagrams and titles are filmed on a special rig which includes a stand to support the camera and the work to be photographed, a means of accurately positioning both and a special camera.

**Stand**

Rigidity and accuracy are the very essence of the design and construction of an animation stand. The camera is positioned, pointing downwards, on one or two vertical columns, usually about 12ft (3.6m) high, and may be moved upward and downward (further from, or closer to, the work) by means of an electric drive, This movement is called tracking or zooming.

A counter indicates the exact height of the camera so that a movement may be accurately reproduced when multiple pre-exposures are necessary.

**Table**

The work, which is usually in the form of a number of transparent sheets known as 'gels', is accurately located in position by a system of pegs and by a glass platten which usually has an aperture of 1ft 2in × 9in and thus holds the work flat.

The pegs are fitted to tracks which may be moved in 1/100 in increments from side to side (east to west) or further away, or closer towards the operator (north and south). The peg tracks may also be rotated or moved on a diagonal (compound) and additional 'floating' peg tracks may be used to locate gels independently.

**Additional accessories**

A pantograph may be used to co-ordinate complicated table movements by means of a pointer.

An aerial-image projector may be used to project an image, from beneath the platten in a mode akin to that of an optical printer. In this manner pre-shot live action may be combined with an animated image.

Computers are frequently used to control and co-ordinate all the various camera and cable movements.

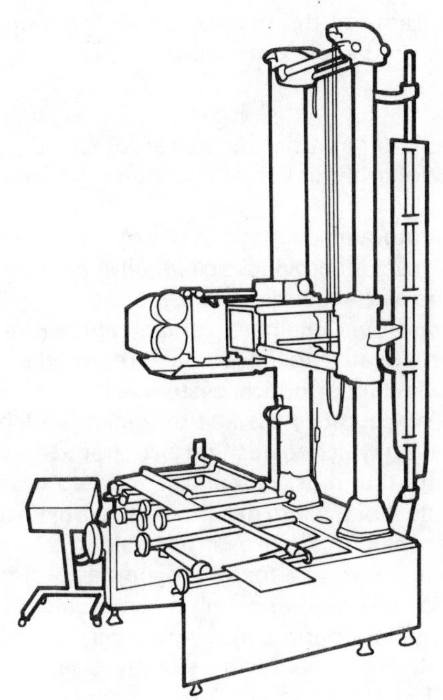

ANIMATION STAND AND TABLE

Oxberry animation stand and compound.

# The Animation Camera

An animation camera incorporates many special features and is capable of a greater degree of accuracy in film registration than a normal motion picture camera, even one fitted with registration pins.

To ensure optimum flat registration for multiple exposure with the camera running either forward or in reverse, animation cameras are usually fitted with 'Bell and Howell' shuttle gate type movements which incorporate permanently fixed pilot pins of exceptional accuracy. The film is advanced from one frame to the next by being lifted off the pilot pins, moved along and lowered again onto the pins.

## Special facilities

The in-shot variable shutter may be adjusted automatically over a period of 6–128 frames either 'logarithmically' for a fade or 'sinusoidally' for a dissolve.

Focus is adjusted automatically, by means of a cam, as the camera is raised and lowered. A cam lift provides an overall adjustment, if the work to be photographed is raised above the table.

Viewfinding is by rack-over, the body containing the film and the movement being slid across, relative to the lens, and replaced by a ground glass viewing screen and viewfinder optical system.

The ground glass is accurately marked to match the film aperture and register pins provided on which to position a cut frame for lining up two or more parts of a picture that must be optically printed together later.

The rear pressure plate of the film gate may be removed and replaced by a light source (rotoscope). This makes it possible to trace a line from an existing scene, as when making a matte for composite printing. With this facility, it is possible to make a series of gels that may later be photographed to form a travelling matte around a moving object.

Immediately beneath the camera is the shadow board. This black board ensures that a reflection of the camera does not show in the glass holding down the subject matter. Other shadow boards, which do reflect, are used to produce special effects.

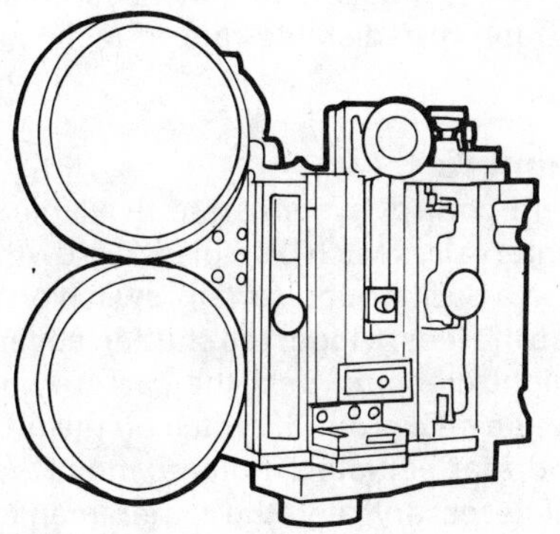



Oxberry 16/35mm process camera.

# Time-Lapse Cinematography

If a film camera runs at less than the standard 24fps projection speed, the action filmed appears to be speeded up. Events, which may have taken hours, weeks, or even months to occur may be filmed at regular intervals over that period and then viewed in a matter of minutes or seconds. Typical uses for time-lapse cinematography are in industrial processes, biological growth, time and motion study, environmental changes, etc.

To do such filming satisfactorily requires a considerable control of the camera, and sometimes even of the subject and its environment. Plants which appear to grow from seedlings to full growth in a matter of seconds must be encouraged to grow with their leaves turned in a consistent direction and not change attitude between dawn and dusk. The soil in which they grow must be kept at a constant state of humidity and not alter colour with watering and, if necessary, the camera may have to tilt upwards with the plant as it grows, at an unpredictable rate of inches per month—a very highly skilled job.

## Control of the camera

Intervalometer camera control systems are available which will, at predetermined regular intervals, switch off a plant's growth light, black out the ambient light, position a background curtain, switch on the filming light (or actuate a flash), uncap the lens, trigger the shutter, advance the film, cap the lens (to stop stray light creeping past the camera shutter), turn off the filming light, remove the background, switch on the growing lights, return the environment to normal, actuate a motor to move the subject or camera between shots and, if necessary, alert the cameraman to any malfunction.

In cases where short but continuous periods of normal running are desired at predetermined intervals, these too can be programmed.

For stop-frame operation the camera, subject and all accessories must be set on particularly rigid supports, the camera must be more light-tight than is normal and it must be fitted with a single shot, short-exposure motor which does not disturb the camera each time it is operated.

An adjustable shutter is an advantage if the exposure must be changed during the course of a long drawn out take.

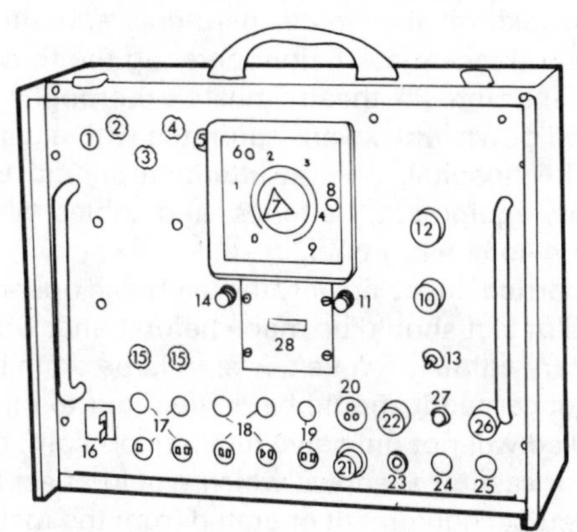


TIME-LAPSE CINEMATOGRAPHY

**Multi-lapse Automatic Camera Control**
1. Mains switch; 2. Lighting fuse (10 amps); 3, 4. Secondary circuit fuses (3 amp); 5. Mains 'on' neon indicator; 6. Fine tuner; 7. Set time interval; 8. Set time range; 9. Range indicator; 10. Test switch; 11. Set to fast time; 12. Shutter release/thrust adjust; 13.Select range – slow or fast; 14. Start timer; 15. Outlet fuses; 16. Mains inlet; 17. Mains outlets and switches; 18. Lighting mains outlets and switches; 19. Remote counter outlet and switch; 20. Switched socket; 21. Curtain winder outlet; 22. Shutter hold time set; 23. Shutter release; 24. Flash charge outlet; 25. Flash trigger; 26. Flash synchronisic control; 27. Flash charge switch; 28. Frame elapsed counter.

*One of the special assignments of a documentary cameraman.*

# Filming an Operation

Films of operations fall into two categories. Those for instructional and record purposes, and those for more general audiences where the emphasis is on the theatre scene and when it is usually preferable not to show too much blood.

## Preparations

A film crew must prepare themselves as well as their equipment and if not used to the sight of major surgery should arrange to watch another operation beforehand. Personal cleanliness is very important. They will probably be required to take off their outer garments and shoes and don a surgical gown, cap and face mask before entering the theatre.

Equipment to be taken into the theatre must be thoroughly cleansed. The exteriors are washed down with swabs saturated with a disinfectant solution provided by the hospital. Special attention should be paid to the undersides of cases, equipment, batteries, and cables which may have been polluted on previous assignments.

If filming is to be carried out using only the overhead operating table light for illumination, a film test should be made beforehand. Optimum results call for supplementary lighting. While this should be done in consultation with the surgeon concerned it should be pointed out to him that a single spot light source used will not only give him a good light, but discourage him and others from casting shadows which would affect the filming.

For safety, all electrical equipment operated from the mains power supply must be properly earthed (grounded), and cables which cross the floor must be taped over to eliminate the risk of tripping. Cables which go up a wall and are likely to be 'in-shot' may also be taped over to make them less unsightly.

## Shooting

If required to shoot the surgical details of an operation deep inside the patient, the cameraman may find difficulty in getting a high enough viewpoint. He must either be on an elevated platform, or a very tall tripod or dolly (be careful that the whole lot does not fall over), shoot through a mirror placed above the patient (everything is reversed, but may be corrected in the printing) or through an inclining prism set 'upside-down' for a high angle point of view. The best position is usually over the surgeon's left shoulder.

There will be no problem in achieving lip sync with the sound as all lips are hidden behind surgical masks. A greater problem is that of matching particular dialogue to the correct pieces of film—especially important if the surgeon gives a running commentary. This can be overcome by using a single system recording, even just as a guide track.

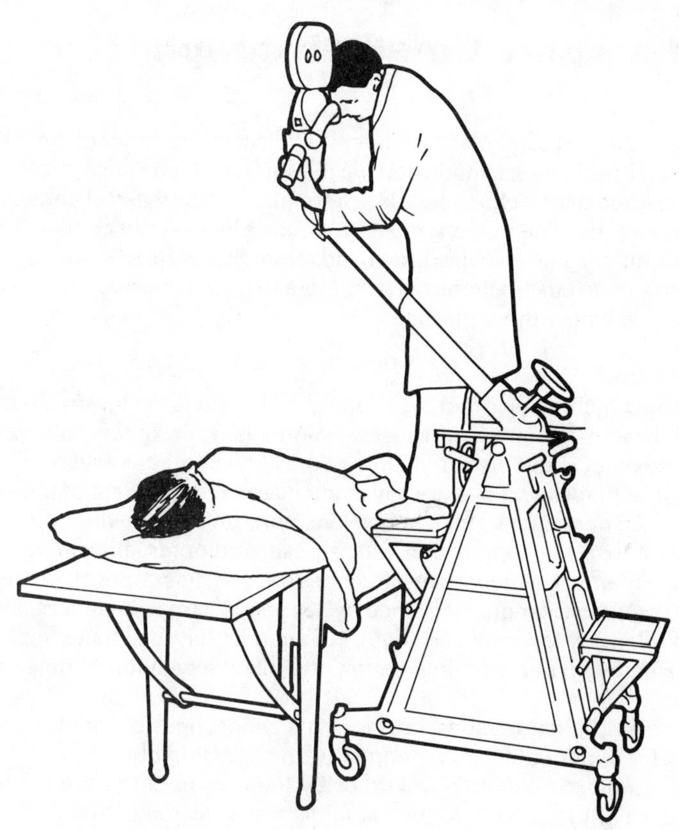

FILMING AN OPERATION

A special camera support used to position a camera directly above a patient on an operating table.

# Underwater Cinematography

The use of the film camera underwater is subject to a number of limitations that do not apply when shooting above the surface. Scuba diving is an art in itself and good diving experience is a prerequisite of successful underwater cinematography. The camera must be housed in a sealed container with windows for the lens and viewfinder and remotely operated controls which permit the principal functions of the camera to be adjusted while the entire ensemble is below the surface.

## Refraction

Apart from anything else that may happen to it, light is bent (refracted) as it passes through water. This causes astigmatism, chromatic aberration, coma and pincushion distortion. But as the subject matter is very often of unrelatable shapes and colours, few audiences are ever aware of the short-comings. A concentric dome-shaped lens port, rather than the normal flat glass front, used in conjunction with a positive diopter lens between the camera lens and port reduces or eliminates most of these optical problems.

Images underwater are magnified by 33⅓ per cent, making lenses appear to be of relatively longer focal length. Put another way, distances appear to be reduced by 25 per cent. In addition, the relative position of objects are changed.

For this, and other reasons, lenses of the widest possible angles should be used for underwater cinematography and focusing done by eye, either through a reflex viewfinder system or by the judgement of the operator, where eyesight is subject to the same distortion as the camera. If a tape measure were to be used the lens would have to be specially calibrated with each foot measuring only 9in (1m measuring 75cm).

## Waterproofing

Perhaps the greatest problem in underwater cinematography is the ever-present possibility that the housing may leak and the camera be severely damaged.

All housings have a depth limit to which they may safely be operated and if the operator is in any doubt the housing should be first lowered to the working depth, or below, *without* the camera installed.

As a precaution against leakage, even when working just below the surface, many housings have provision for the inside to be inflated with a tyre pump to a pressure of 10–12lbs psi. If this is done, any leak will be indicated by a stream of air bubbles which act as a warning to the operator to return the housing to the surface immediately.

Should the camera be contaminated by sea water emergency cleansing treatment must be carried out as soon as possible.

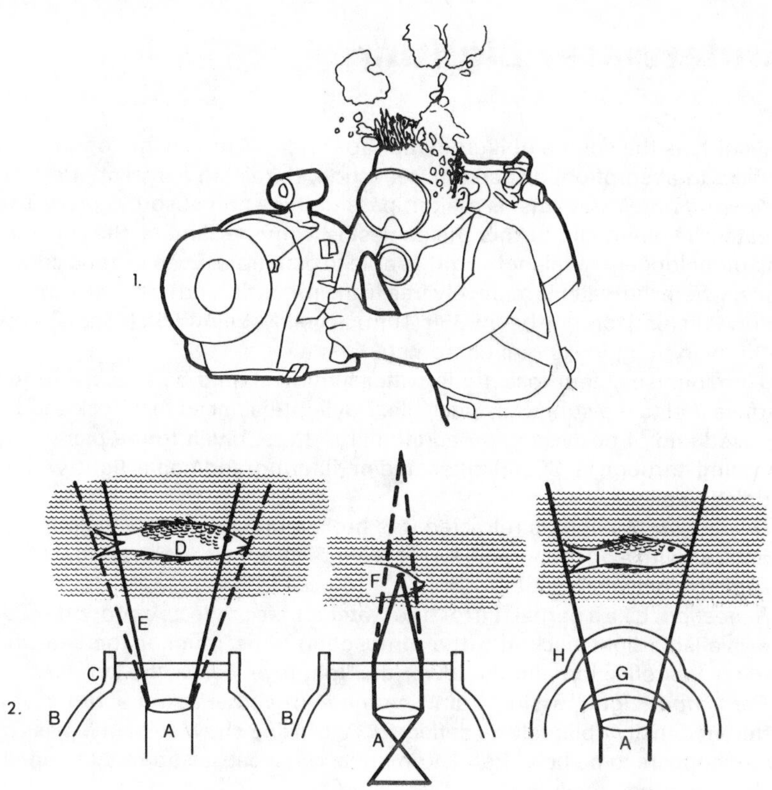


UNDERWATER CINEMATOGRAPHY

1. Underwater cameraman with camera; 2: A. Camera lens; B. Waterproof camera housing; C. Flat glass port; D. Object as seen in air; E. Lens angle is reduced by refraction underwater making objects appear closer to the camera than they really are; F. Lens must be focused on apparent distance not on real distance according to focus scaling appropriate to surface use; G. Diopter lens; H. Concentric domed lens port; I. With G and H object appears to be correct size and suffering less distortion than with a flat front.

# Underwater Lighting

Daylight, as the source of illumination for underwater cinematography is subject to absorption, dispersion, reflection, refraction and scattering.

*Absorption*. Water acts as a filter, passing blue and absorbing red. The greater the depth and further the subject is from the camera, the less red, and other longer-wavelength light, is able to penetrate. Even in good conditions, where the water is relatively free from impurities, no red light remains below 20ft (6m), orange below 35ft (10m) or yellow below 65ft (20m). Below 100ft everything looks monochromatic.

To minimise colour loss, underwater filming should be as close to the surface and to the subject as possible. Daylight-balanced filmstock should be used and, if necessary, an additional red cast given to the picture by shooting through a CC20R filter and/or filtering additional light with a reddish filter.

*Dispersion*. As light is refracted it is broken into its spectral colour elements, as with a prism or rainbow. This aversely affects focus, contrast and distinction between colours.

*Reflection*. Like a partial mirror, the water surface reflects a proportion of the available light back into the atmosphere. The rougher the sea, the greater the reflection and the lower the light level below the surface.

*Refraction*. Light passing from one medium and entering another of a different density obliquely, is deflected. A drinking straw seen in a glass of water appears to be bent. Refraction underwater causes objects to appear closer than they really are.

*Scatter*. All but pure distilled water contains impurities and even this when aeriated becomes almost opaque as the quantity of air bubbles increases. Sea water contains suspended impurities, minute sea animals, plankton and air bubbles. All combine to reduce both the intensity of the light and the contrast and colour saturation of any object seen through it.

### Supplementary lighting

Battery-powered tungsten halogen lights in watertight containers may be used underwater. They should be placed between 45° and 90° from the subject relative to the camera to reduce the amount of light reflected back into the lens by the suspended particles and air bubbles in the water. Underwater lights are designed to be used underwater only and overheat if switched on above the surface. If fitted with vent plugs it is important to check that these are opened during charging and firmly sealed when the lamp is submerged.

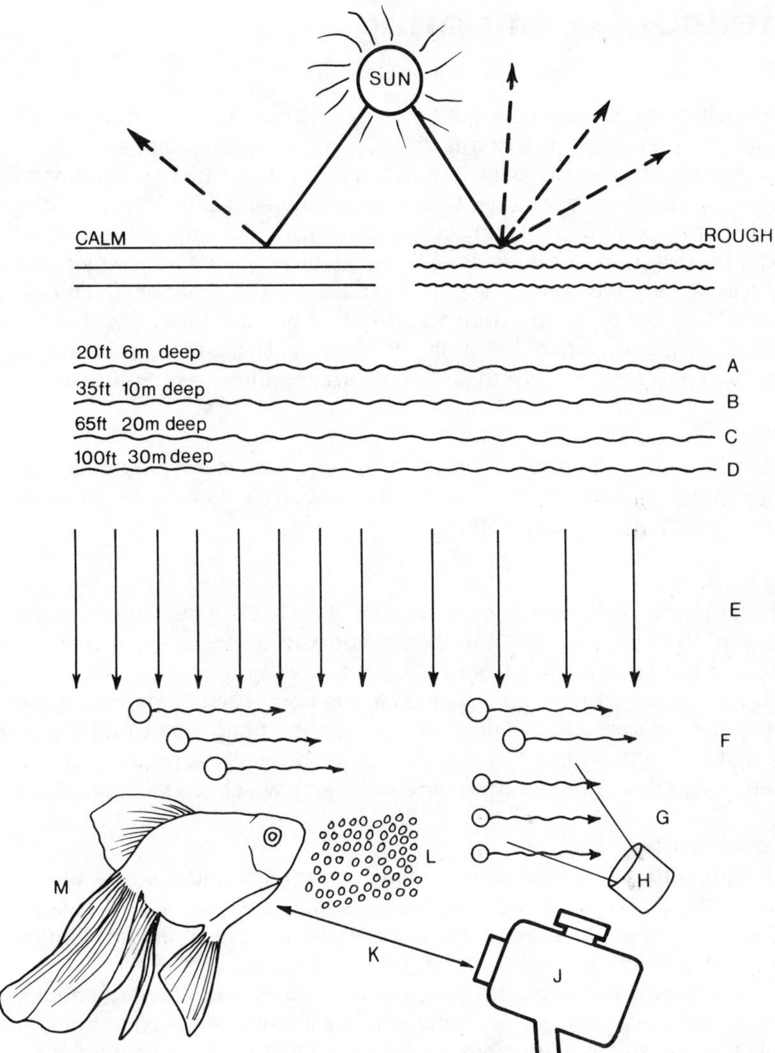


UNDERWATER LIGHTING

**Conditions affecting underwater illumination**
A. No more red light; B. No more orange light; C. No more yellow light; D. Everything black and white; E. Less light passes through the rough surface; F. Sunlight scattered from particles of water; G. Light scatter from artificial light source (H); J. Underwater camera, light is reflected from subject absorbed and scattered by particles and air bubbles (L). Distance (K) from the camera to the subject (M) should be as little as possible.

# Helicopter Shooting

The helicopter is a popular, but basically unstable and frustrating camera platform. It vibrates, it changes its level every time it changes its speed, there is a limit to how close it may approach a subject, there is a temptation to use lenses which are longer than prudent, there is a difficulty in holding a constant distance from a subject for purposes of critically focusing a long lens. Focusing may only be done by eye, which may be difficult when the lens is well stopped down. The rotor blades cause a draught which ruffles artists' hair and blows up surface dust and the blades cause a shadow in the top of the picture when the camera is tilted up. Helicopters are expensive and dangerous, but, nevertheless, they are the film-maker's dream.

## Pilot

The most vital requirement for helicopter shooting is a highly skilled pilot who appreciates the problems of the cameraman, a man, in fact, who flies a camera rather than a helicopter.

## Helicopter

The larger the helicopter, the steadier it is likely to be. It must be in a perfect state of maintenance and the blades checked to ensure that they pass through the same plane while rotating at operating speeds. This is done by rubbing different coloured chalks on to the tip of each blade, running the helicopter up to speed and positioning a leather thong on the end of a pole so that it is lightly rubbed by the blade tips. The chalk marks show if all the blades pass through the same plane and, if not, which are failing to do so.

## Cameraman

For optimum results, the cameraman must understand the limitations of the helicopter. A helicopter achieves changes in direction and speed by altering the angle of attack of the rotor blades relative to the fuselage. For greatest stability, the helicopter must be flying forward.

Shots in which the camera appears to move directly forward should, if possible, be avoided as this involves side-slipping the helicopter in flight. This is a less stable manoeuvre than forward flight due to the wind buffetting against the comparatively unaerodynamic side of the helicopter.

No cameraman should fly in a helicopter until he has personally checked that suitable insurance cover has been taken out and paid for.

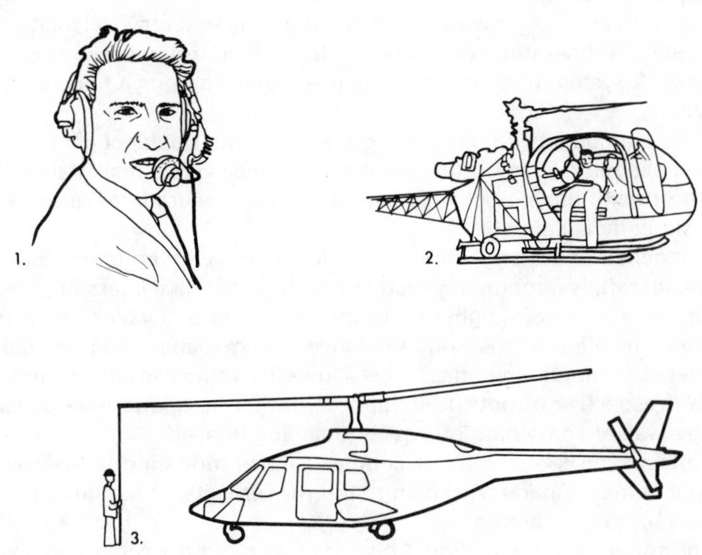

HELICOPTER SHOOTING

1. Some pilots are better than others; 2. Pilot and cameraman must work as a close-knit team; 3. Check blade tracking to ensure minimum vibrations.

# Helicopter Mount

Helicopter filming begins where the dolly or crane-mounted camera leaves off. From a helicopter it is possible to start a scene with an artist's face filling the screen (shot on the long end of a 10:1 zoom from six to ten ft away) and then widen the lens angle as the helicopter pulls away until the original subject is an indistinguishable speck on the horizon.

## Helicopter mount

When using really long lenses in a helicopter it is essential to isolate the camera and the operator from the short high frequency vibrations of the helicopter. A system of spring mounting is usually employed and is very effective.

The cameraman sits on this sprung platform with no direct connection with the helicopter. Even his feet must be on a special foot rest attached to the platform and the straps by which he must be secured, for reasons of safety, similarly isolated.

Also attached to the anti-vibration platform, and with no direct contact with the aircraft, is a mounting system which holds the camera in perfect balance, against mass weight, about the pitch, roll and yaw axes of the helicopter. The effect of this counterbalance system of mounting is to make the camera seemingly weightless. This allows it to remain in any position or attitude irrespective of outside influences, rather as a long pole gives a tightrope walker something by which to steady himself.

The effectiveness of the mass counterbalance mounting is destroyed when the camera operator transmits the movements which he is being subjected to, to the camera.

When properly set up, it should be possible to put the camera in absolutely any position where it should remain perfectly still. So sensitive is the state of balance that it is necessary to make slight adjustments to the trim of the mount as film is passed from one compartment of the magazine to the other.

Lightweight helicopter mounts intended primarily for 16mm usage in small helicopters do not isolate the cameraman and the camera to the same extent as is possible with the large mounts. Consequently, extremely long lenses cannot be used without showing unsteadiness.

## Operating

The secret of good helicopter operating is to let the mount do its work and to interfere as little as possible. If it were possible to aim the camera only by blowing onto it, this would be ideal. As it is, light fingertip contact only should be used. For this reason focus and zoom operation must be done by remote electric control. To grip the lens would undoubtedly transmit unwanted movement to the camera.

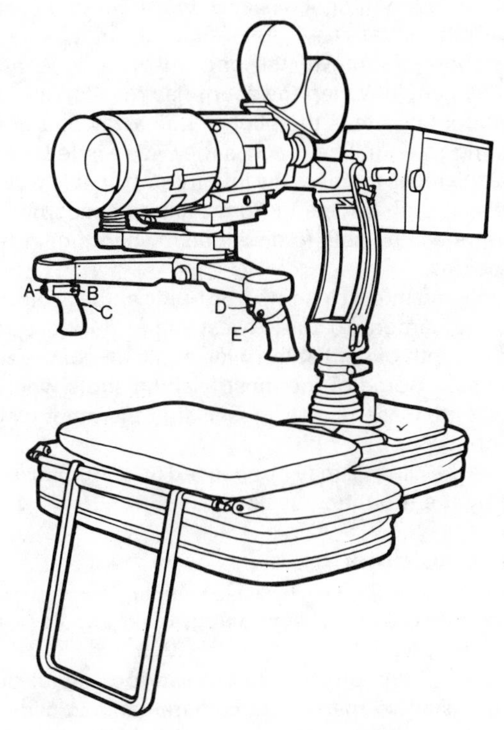


HELICOPTER MOUNTING

A. Tyler 'Major' helicopter mount.
Operating controls: A. Focus closer; B. Camera start and stop; C. Zoom out; D. Focus to infinity; G. Zoom in.

# Planning Helicopter Shots

When planning a shot, the cameraman and pilot must co-operate in positioning the helicopter, bearing in mind that a helicopter may only hover steadily by flying into the wind. A full discussion should take place and the plan of action discussed before take-off.

When shooting a scene which starts with a close-up and progresses to a wide angle, the cameraman uses the aircraft intercom to tell the pilot where to position, sets the zoom lens to the long focal length and focuses up on the subject. As soon as everything looks as if it is about to be correct he must start shooting, for the situation will not remain like that for long. As soon as he has sufficient footage shot with the long end of the zoom he must start to zoom back to wide angle. When the zoom has reached about half-way to maximum wide angle he tells the pilot to pull away and gradually slows down the zooming rate until he reaches the wide angle stop. At the same time, he adjusts the focus towards the infinity stop. In this way, it is possible to get a perfectly smooth transition from zoom to helicopter movement in which few viewers will be able to distinguish where one movement ends and the other begins.

To do a shot moving into a close-up from wide angle is more difficult. It is usual to focus the camera to the closest range during a rehearsal and depend upon the depth of field at the wide angle end of the zoom range to take care of the rest. Some of the most exciting shots where the camera appears to move into close-up are, in fact, shot the other way around and reversed in printing.

Where possible the camera may be speeded up to minimise any vibration not eliminated by the mounting system.

### Don't be too ambitious

Incredible helicopter shots are filmed by very few master aerial film cameramen. The inexperienced cinematographer should confine himself to wide angle coverage until he is familiar with the results. What may look good through a long lens on the camera can look awful on the screen. Helicopter filming is a two-man effort, cameraman and pilot. The cameraman should chat to the pilot on the intercom to build up a rapport so that when the camera is running both men are thinking and working together.

PLANNING HELICOPTER SHOTS

Planning Nelson Tyler's 'Statue of Liberty' shot for Kodak's famous TV commercial. The shot was photographed as shown and then printed in reverse giving the effect of moving into close-up, and the man pressing the camera button in the last second of the shot.
1. Start shot when all is steady; 2. Start zooming back as finger is raised; 3. Helicopter starts pulling away; 4. Zoom complete; 5. Helicopter continued to pull away; 6. And circle to the left; 7. Helicopter reaches altitude of 300ft.

# Aerial Cinematography

Air-to-air or air-to-ground shooting from a fixed-wing aircraft is less difficult than shooting close up from a helicopter. Nevertheless, to achieve the best results, the conditions must be right and patience is essential.

## Safety first

If the aircraft is large and the filming is to be done through an open hatch, the cameraman should wear a back-pack parachute all the time. He should be attached to the aircraft by a long strap which is not quite long enough to allow him to fall out and if the flying is to be over water, he must also wear an inflatable life-jacket. Cameramen have been killed because they did not observe these essential safety rules and thought it 'would not happen to them'.

When filming from an aircraft fitted with ejector seats, the cameraman should ensure that in the case of an emergency, none of his equipment would impede his rapid egress from the aircraft. This includes the battery-to-camera cable which, if the battery is secured to the body of the aircraft, should only be lightly plugged in so that it may pull away easily if it is necessary to jettison the camera.

As with helicopter shooting, no cameraman should fly in a non-scheduled aircraft until he is sure that full insurance cover has been taken out both for himself and the equipment, and paid for.

## Filming practice

Large fixed-wing aircraft often fly so smoothly that a cine camera with a short focal length lens does not need any form of stabilisation. A tripod with a levelling head should be securely fitted by an engineer familiar with aircraft regulations.

Cameras are often operated at 32–48fps or faster to smooth out vibrations and slow down relative movements. With long lenses, an anti-vibration mounting system or an image stabiliser is needed for steady high definition pictures.

For hand-held work in small aircraft, shoulder-supported cameras are best, especially in jet fighters or when filming aerobatics air-to-air, when '4g' forces make a camera and operator weigh four times their normal weight.

A two-way intercom between cameraman and pilot is essential. An 85B filter reduces the blueness of high altitudes and a pola screen improves contrast if the aircraft is flying constantly in one direction.

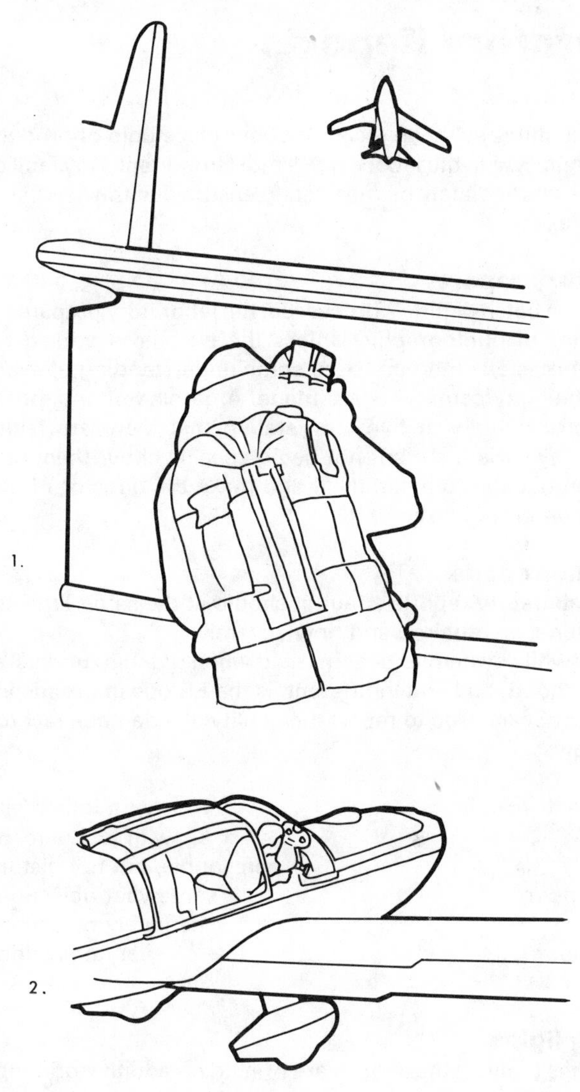


AERIAL CINEMATOGRAPHY

1. Air to air; 2. Air to ground.

*The laboratories' opinion of the cameraman's work.*

# Laboratory Reports

Whether a unit is away on location or out of a studio or production office, regular contact with the laboratory is most important. Any fault or inconsistency must immediately be reported to ensure that the necessary remedial action is taken.

## Laboratory reports

Every time a batch of film is processed, the laboratory prepares a report for the director of photography. Unless the two have worked together on previous occasions there must be some understanding of exactly what is meant when any comments are made. A garbled phone message to the production secretary at five a.m. stating that there are faults with the previous day's material, but not being specific about them, is enough to bring a production to a standstill and make the director of photography *persona non grata* on the unit.

## Impairment code

When a laboratory reports a fault it should at the same time state exactly what material is impaired and how severely.

A worthwhile system to bear in mind when trying to understand what is meant by 'good, bad and indifferent' is the European Broadcasting Union Impairment Scale used to report the quality of videotape recording, which is as follows:

| *Grade* | | *Impairment* |
|---|---|---|
| 1 | Excellent | with imperceptible faults |
| 2 | Good | with just perceptible faults |
| 3 | Fair | perceptible, but not disturbing faults |
| 4 | Fairly poor | somewhat objectionable faults |
| 5 | Poor | definitely objectionable faults |
| 6 | Very poor | very objectionable faults |

## Printing lights

Most rushes (dailies) these days are printed on additive printing machines with the three primary colours controlled individually. A scene has three printing lights, usually in the order red—green—blue. The centre lights on the scale are 25–25–25 but there are inevitably variations across the scale even on a correctly exposed negative, due to variations inherent in the filmstock.

For normal shooting, cameramen should aim to produce a negative which prints on the centre of the green scale.

Ten printing points are roughly equal to one stop of camera exposure.

BATH ROAD, HARMONDSWORTH
WEST DRAYTON, MIDDLESEX
ENGLAND
TEL. No. 01-759 5432
TELEX 22344
CABLES: TECHNICOLOR, WEST DRAYTON

**Technicolor®**

RUSH PRINT

VIEWING REPORT

Sheet No.

CAMERA MAN

Stock Edge No.

COMPANY Fortune Productions Limited

PRODUCTION "THE NIGHT OF FIRE" DATE OF PHOTOGRAPHY 14.4.78.

CAMERAMAN P. Marshall CAMERA SHEET No. 49204 LAB ROLL No. 61100

EMULSION No. 5247-677 COL/B&W R B. CAN GPO No.

| SCENE | TAKE | R | G | B | R | G | B | SCENE DETAILS | EXP. | REMARKS |
|---|---|---|---|---|---|---|---|---|---|---|
| 1 | 1 | 28 | 27 | 31 | | | | DAY EXT 85F | OK | |
| | 2 | - | - | - | | | | 85F ¼ FOG | / | |
| 2 | 4 | 47 | 49 | 50 | | | | | | OVER EXPOSED - Printing |
| | 5 | / | / | / | | | | | | Top of Scale |
| 3 | 3 | 27 | 29 | 32 | | | | | OK | |
| | 4 | - | - | / | | | | | / | |
| 4 | 1 | 29 | 30 | 31 | | | | DAY INT | OK | Slightly forward focus |
| | 3 | 28 | 29 | 30 | | | | | / | OK |
| | 4 | / | / | / | | | | | | |
| 5 | 3 | 30 | 31 | 31 | | | | | OK | Slight DA Top Centre |
| | 5 | 29 | 30 | 31 | | | | | / | OK |

DELIVERY INSTRUCTIONS

EMI STUDIOS
BOREHAM WOOD

SPECIAL COMMENTS

Note 2-4 } 2-5 } Over Exposed - Printing Top Scale

5-3 Slight DA Top Centre - not serious will be masked off in projection.

LABORATORY CONTACT IS: P.Chambers

8/367

LABORATORY REPORTS

Rush print.

# The Laboratory Connection

Every laboratory employs 'client contact' or 'client liaison' men whose job it is to keep you informed, help you to get better service and to pour you a drink occasionally. Use them.

## Visit the labs

Any client is made most welcome if he asks to see the processing plant, to meet the person who is grading (timing) his film and to have a demonstration of how their latest colour analyser operates.

Provided you give them notice and fix a time and day, almost any laboratory is delighted to give an instant course on sensitometry, colour appreciation, negative handling and print assessment. (If they aren't, they probably have something to hide and are not worth using anyway.) If you are having sufficient film processed they may even time your appointment to coincide with lunch.

Talk to any lab contact man and they all tell you the same thing—they wish that more cameramen would pay them a visit so that they could explain their problems, liaise and work out a means of closer cooperation. When you have a batch of film going in that is particularly important to you or in some way difficult, tell them. If you cannot reach the client contact man, phone the supervisor. Ask the lab to phone you too, even if it is in the middle of the night, if they have a query for you. Give them your home or hotel phone number and take theirs as well.

When away on location, very often the most reliable form of communication is by telex. All laboratories have telex machines. Find out if your hotel has a telex or find someone locally who will pass on messages and ensure that important messages are sent, or confirmed by this means.

## He who pays the piper

Remember, well exposed film is useless unless it is handled properly in the labs, and labs are useless without clients. In return, directors of photography and all other specialist film cameramen should take the opportunity, when they have an interesting sequence to film, of inviting the lab contact man and even the supervisor to visit the set or location. Such interchange can only help.

**Hazeltine colour film analyser**
One of the many items of laboratory equipment which the laboratory contact man will be pleased to show and explain to any cameraman who shows interest in how his film is processed.

# Viewing Rushes (Dailies)

The regular viewing of rushes (dailies) is important for all the senior people most closely involved with a production.

## Graded (timed) or ungraded (one light)?

Rushes (dailies) are less expensive if printed 'ungraded' or 'one light'. For the director of photography such prints show up any errors in exposure of colour balance which may be corrected later but are discouraging to non-technical people who see them. Particularly in the field of TV commercials, where the representatives of the advertising agency and even the manufacturer of the product might be present, it is psychologically advantageous to order graded colour rushes (and to supply the cameraman with a list of all the printing lights and colour corrections).

Where time and budgets permit, the laboratories prefer to supply graded colour rushes because it makes their processing look better and they can charge more.

## Viewing rushes (dailies) on location

When working in a studio or a film production centre, there are always facilities for viewing rushes (dailies) under good conditions, but on location it is often a different matter.

Rushes may be shown at a local cinema after the end of the viewing performance but this is not ideal. Apart from the fact that some cinemas have poor projection facilities with dirty lenses, ports and screens and are not equipped for double head (sep-mag) sound viewing, sessions which commence after 11pm keep everyone up late at night. Directors of photography, bearing heavy responsibilities, invariably prefer to go to bed reasonably early.

The satisfactory alternative is to take away on location a portable double head Xenon light projector. Such machines can show a bright picture up to 12ft wide, with reasonable sound so that everyone concerned will know exactly how the production is progressing. If, in addition, the rough cut is also being done on location, those concerned will also be able to see the film assembled to date.

# LOCATION VIEWING AND CHECKING

1. Single head portable projector can show mute (silent) print or com-opt.

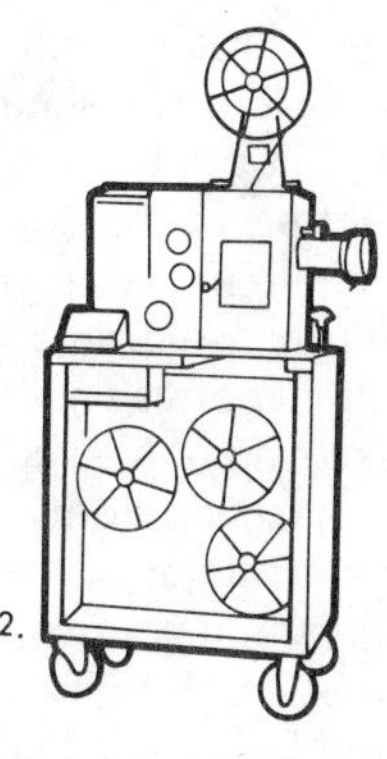


2. Double head portable projector can show mute (silent) print, sep-mag or com-opt.

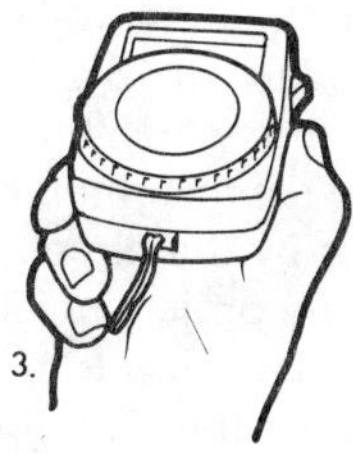


3. An ordinary reflected light exposure meter can be used to measure screen brightness. Check the screen brightness of one or more reputable laboratory screening rooms to set a 'standard'.

# The Answer Print

Each one of the many hundreds of scenes which were shot and are now cut and intercut to form a motion picture must be graded (timed) for density and colour balance.

Film is a creative art and as such the interpretation of how the finished product should look visually cannot always be clearly defined. The person who should be the arbiter of what is desirable is the man whose name appears on the credits as the director of photography.

After a film has completed the editing and sound dubbing stages, the negative is cut to match the cutting copy and an 'answer print' made. Inevitably, some of the scenes in the first print will be too light or dark or too red or green. The permutations are endless. Any comments by the producer, director and the director of photography are noted by the laboratory and translated into a second answer print—and so on until a near perfect print is achieved.

Very often at this stage, a colour intermediate positive and a colour intermediate negative or a 'CRI' (Colour Reversal Internegative) are made and all release prints made there or elsewhere are taken from the duplicate negative with little need for further grading (timing). That leaves the master negative secure from possible damage. A new answer print is made at each lab from the duplicate negative to make sure that their prints will be satisfactory.

It angers many directors of photography that sections of the bad answer prints made before the ideal was achieved are retained and subsequently distributed to provincial cinemas. This undesirable practice is, regrettably, an economic necessity.

## Screen brightness

Screen illumination at the laboratory is most likely to conform to the laid down standard of 16 $\pm$ 2 Foot Lamberts. The same cannot be said for all the cinemas in which the film is eventually shown. Inevitably, some will be darker and some lighter. From here on all the care and dedication lavished on the film by the director of photography and the camera crew is in the hands of the projectionists at local cinemas all over the world.

With a major film to be premiered in an important cinema it might be advisable for the director of photography to visit the cinema before it opens one morning and ask if he may measure the screen brightness. This he can do with his reflected light exposure meter, and compare it with a similar reading off the screen at the lab. There are occasions when a special print, either darker or lighter than the normal, has been made to ensure that during its 'show-case presentation' the film should look exactly as those who made it want it to look.

THE ANSWER PRINT

The *'coup de grace'*, and final involvement of the director of photograpy in a film production is to approve the perfect print.

# The Camera Team

Cinematographers vary widely in their personal specialisations from cinema feature work to newsreel, from the traditional to the *avant garde*. There are those whose forte is the 'creative' style, those who are interested in the technical aspects, those who have a bent for journalism, those who 'think big' and those who prefer the intimate. No two cameramen are exactly the same and many who can be considered to be among the best within their particular field would be lost and utterly wrong for a picture which is not within their range of experience.

In effect, cameramen must be cast for a particular assignment just as actors and actresses are cast to play different parts.

## Camera crew

How many technicians are employed to form the camera crew depends upon the type of filming, the type of camera equipment to be used and, very often, upon local trade union agreements.

Usually the most efficient and economical way to go about shooting a feature film using a studio-type camera (whether the picture is being shot in a studio, or on location) is to employ a four-man crew—director of photography, operator, focus assistant, and clapper loader. The grip completes the team. Eliminate any one and, if it increases the time taken, the cost of shooting a feature film must inevitably escalate disproportionately. If it introduces errors, the quality declines.

Second unit exterior photography without principal artists, documentary TV, advertising, commercial, newsreel and scientific cinematography, etc all have different requirements. Sometimes three, two or one man on his own, doing all the camera jobs plus a lot more besides are not only adequate but more sensible.

It all depends on the circumstances and the equipment.

## Be nice

Where more than one constitutes a camera crew, or even a film unit, personality and the ability to mix and live with others plays a significant role in whether or not a person makes a successful career in the film industry. Film making is essentially teamwork and nice people work best if they have nice people around them. There is no place for the abrasive loner. There is an old saying in the industry which goes 'Be nice to those you pass on the way up, you may meet them again on the way down.'

THE FEATURE CAMERA CREW

A. Director of photography; B. Grip; C. Operator; D. Focus/1st Assistant; E. Clapper loader/2nd Assistant.

*An artist as well as a technician.*

# Director of Photography

Because the amounts of money involved are so vast, no-one is likely to be appointed director of photography unless he has proved both his artistic ability and his sense of responsibility. No matter what his background, he will probably have to serve an apprenticeship starting as a clapper loader.

When he has proved that he can carry out these duties in a responsible manner, he will get a break to pull focus, probably on a second unit, and when he has proved himself there, to first unit. Similarly he will get an opportunity to operate and eventually to light, at each step gaining experience by working on different types of films and with different people while at the same time proving his reliability.

Some climb the ladder faster than others, some choose to make a career out of being a very good focus assistant or operator, some find they have a particular talent and become specialists.

## Working relationships

The director of photography works closely with the director and the production designer of the film. Between them they decide upon the mood it is wished to achieve for any particular sequence, where the camera should be set up for particular shots, the angle of the lens to be used, and what movements the camera should make during the course of the shot.

While the rest of the crew physically handle the equipment, his prime concern is with realising the desired 'look' of the scene and, in effect, painting with light. For his palette he uses lamps, possibly with coloured filters, nets, gauzes, diffusers, reflectors and the means of masking off unwanted light. With an exposure meter, and perhaps a colour temperature meter he measures and assesses the lighting.

He has further control in the photographic filters, diffusers, fog filters, grads, nets, pola screens, star filters and other front-of lens effects, in how he sets the lens aperture, how much he reduces the shutter opening and in the control which he asks the laboratory to exercise in the processing and printing of the film.

In his lighting he must be concerned with the continuity of effect that he may be creating so that a number of scenes may be edited together to form a cohesive sequence, even though shot out of sequence, on location, in a studio, wide-angle and close-up and from many different angles.

Ageing actresses expect him to lose some of their wrinkles, balding actors will expect kindness shown to their pates, children must be humoured and animals not frightened—all part of his accumulated skill.

## THE DIRECTOR OF PHOTOGRAPHY'S MEANS OF IMAGE CONTROL

| | |
|---|---|
| Film size | Super 8, 16mm, 35mm |
| Picture shape | Academy, wide screen, 'scope |
| Camera support | tripod, dolly or crane mounted, hand-held, floating-camera system, airborne, underwater |
| Camera movement | in-shot panning, tilting, tracking, dollying, rise and fall |
| Camera position | relative to subject, relative to the ground and other fixed objects, orientation |
| Camera speed | normal, speeded-up, slowed down |
| Camera shutter | fixed, adjustable stationary, adjustable in shot |
| Lens focal length | wide angle, normal, telephoto, zoom |
| Lens point of focus | for a definite plane or for depth of field |
| Lens aperture | normal, wide aperture |
| Exposure | normal, under- or overexposure |
| Lens accessories | close focusing, range extenders, diopters, prisms |
| Filters | colour temperature control, colour biasing, neutral density, graduated, polarising, definition-degrading, low contrast, fog effect, star effect, nets |
| Light | natural, artificial, reflected, augmented, controlled |
| Light control | intensity, hardness, diffusion, colour, highlights, shadows, reflections |
| Filmstock | type, exposure, flashing, modified processing |
| Print | stock type, colour grading (timing), density grading (timing) |

# The Operator

The operator is responsible for the composition of the picture as it appears on the screen. As with the director of photography, he is able to use a fair amount of artistic licence if it is in accordance with the wishes of the director of the film.

Good operating, because it is good, is rarely noticeable to the ordinary cinemagoer. Bad operating destroys the illusion by subconsciously informing the viewer that he is watching a contrived situation. Uneven pans and tilts, follow-shots where the moving object does not remain at a constant distance from the edge of the screen, bad composition, tops of heads cut off, awkward cuts across people's legs, these and many others are the cardinal sins of bad operating.

The mounting and the viewfinder system are the operator's instruments over which his hands and eye, in complete coordination, have control.

## Final decision

After a scene is shot, even if the director is satisfied that the take is good enough to print, the operator is asked if it was acceptable for him. Until the film has been processed and subsequently screened no one but he sees a scene as it was actually recorded. After processing, it may be too late or far too expensive to re-shoot. It is possible for the director to make his own final decision only if a TV viewfinder system is used.

If a camera, and the crew, are mounted on a moving platform such as a dolley or a crane, the operator requires close co-operation from the grip who pushes the dolley or crane when 'in-shot' tracking or jibbing movements are required.

## Tools of the trade

The operator must have a choice of cameras, heads and mountings available to him, according to the requirements of the picture. The studio camera and the lightweight hand-held type camera are complementary and the one does not do the job of the other. Only the very latest 35mm cameras (eg: Arriflex 35BL and Panavision Panaflex) may be considered universal and suitable for use in almost any situation.

Most operators prefer to use a geared head for the most delicate and positive camera movements. Operating a geared head is a skill acquired while working as a clapper loader and focus assistant. After a while, with the frequent need to move the camera while at the same time making technical adjustments it becomes unnecessary to think which way to wind a handle to get a certain camera movement—a reason why geared heads are not favoured by those without the necessary experience. For rapid movements, or when operating the focus or zoom himself, an operator usually prefers a 'free' friction or fluid head.

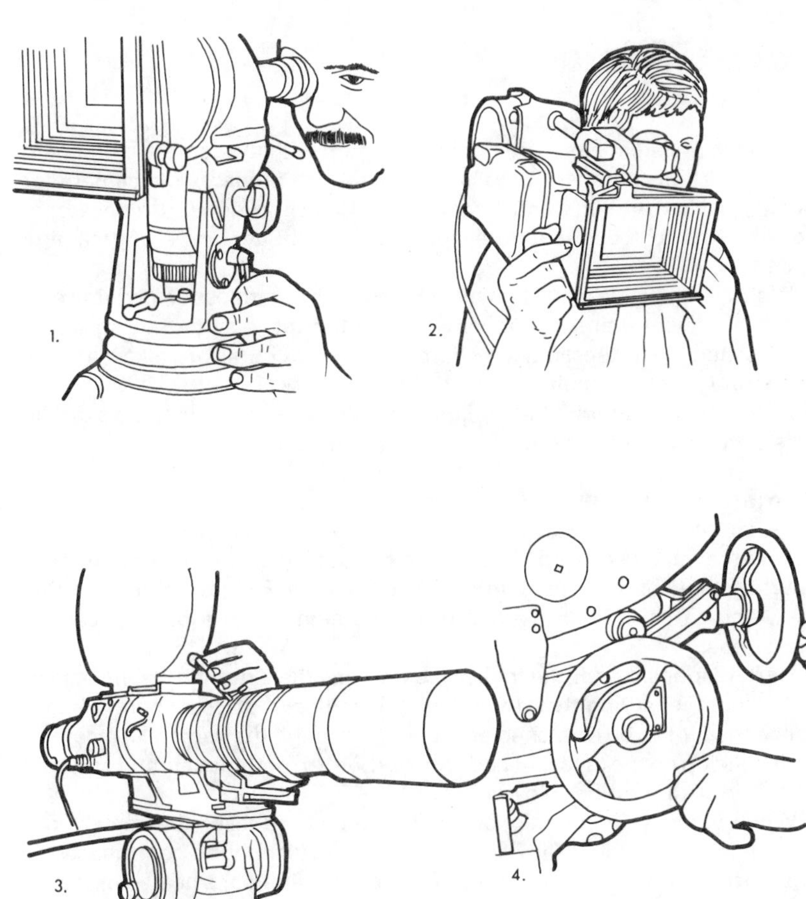


HANDLING THE CAMERA

1. **Friction head**
Used principally where fast or coarse camera movements are required, or where a simple lightweight head is adequate.

2. **Hand-held camera**
Used where freedom of movement is more important than camera steadiness.

3. **Fluid head**
Used for smoother pans and tilts especially with zoom and telephoto lenses. Camera operator has a hand free to focus or zoom if required.

4. **Geared head**
Used for the finest and most accurate camera movements.

*His prime concern is the camera.*

# Focus Assistant

The focus assistant has a greater involvement with the camera equipment than any other member of the camera crew. He prepares the camera for operation, fits the lens and filters needed and makes all the necessary adjustments to focus, aperture, focal length, shutter angle and camera speed.

If there is no single plane that must be in focus regardless of anything else within the picture area, he must calculate the focus setting to make the nearest and the furthest objects sharp. During the course of a take it may be necessary for him to 'pull focus' and change the focal length of a zoom lens. He ensures that no bright stray light that might cause a 'flare' falls on the lens or filter from the front on to the filter from behind.

## Camera cleanliness

After shooting the last take of a particular scene, he checks the gate to ensure that no loose hair or piece of dirt has settled on an inside edge of the aperture plate and obtruded into the picture area. Failure to carry out this task, and to call for an extra take if a foreign body is present can be disastrous.

He is responsible for the internal (as well as the external) cleanliness of the camera, ensuring that no dust or grit settles in places where it might cause the film to be scratched as it passes through the camera. Any emulsion build up on the gate must be removed with a toothpick etc and any dust removed with a rubber blower.

When working in conditions of extreme cold he takes the camera out early to run it, without film, for a period before shooting commences to ensure that it is working smoothly. If heaters are fitted or a heated barney is to be used he must ensure that they are switched on and operating. In conditions of excessive heat he ensures that the film magazines are protected from the direct rays of the sun.

When it rains, he is responsible for ensuring that the camera is adequately covered and dried out if it becomes wet.

Among the items which he usually has to hand are a tape measure, a depth of field calculator, a pen torch, a blower, cleaning materials, toothpicks, a small kit of tools, and a roll of camera tape with which to mark, secure, seal, repair, modify, make, improve and generally use for every imaginable purpose.

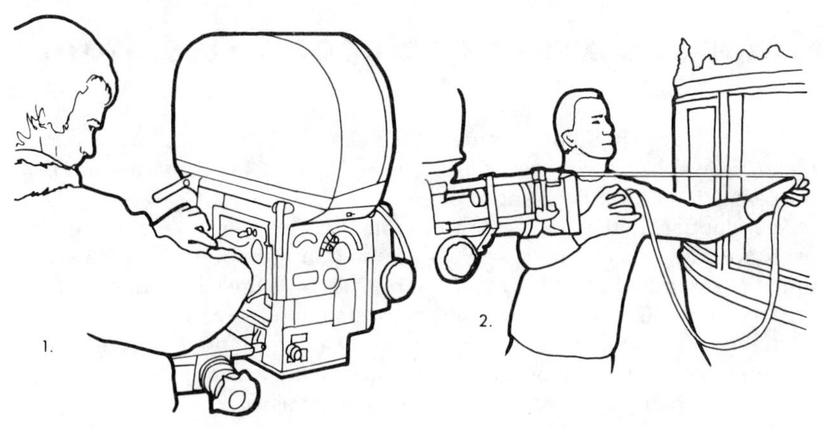


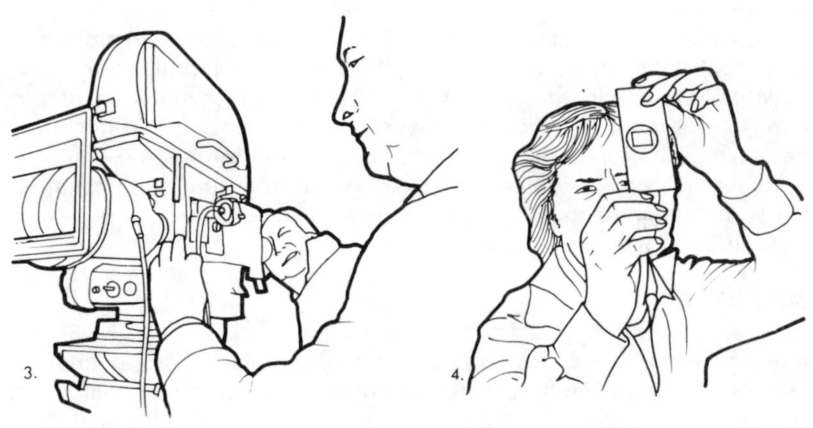


DUTIES OF THE FOCUS ASSISTANT

1. Loading the camera; 2. Measuring focus; 3. Pulling focus; 4. Checking the gate.

# Clapper Loader or Second Assistant

Although the most junior member of the camera crew, the clapper-loader nevertheless, has a very responsible job. One small mistake by him can prove costly in both time and money.

The clapper loader marks and operates the clapper board, helps the assistant as necessary (perhaps by 'zooming' a zoom lens, if at the same time the assistant has a tricky focus to pull) records the length of each take and the contents of each roll, assists in reloading the camera, ascertains from the director which takes are to be marked 'Print' on the report sheet and makes sure that the product of the day's work is safely despatched to the laboratory. The ability to write legibly is an asset.

The clapper loader is also responsible for loading and unloading the film magazine, an operation that must be carried out in conditions of absolute darkness. If he were to allow even the slightest streak of stray light to fall upon the film during this operation or mislay a can of exposed film the combined efforts of a large number of people during the whole shooting session would be wasted. A single roll of film may contain scenes involving expensive artists and thousands of extras or shots of a set which has been burnt down or blown up. If, or when he makes a mistake, he must report it immediately.

The clapper loader keeps an account of the unexposed film stock and deducts that consumed to date. If supplies run low the production office must be informed in good time so that they may order more, bearing in mind that there are few places in the world where unexposed film stock may be obtained after five o'clock at night or at weekends.

On top of all these duties it is usual for him to ensure that a plentiful supply of refreshments is delivered to the remainder of the camera crew at the appropriate times.

### Ambitions

The clapper loader's pockets usually contain pens, pencils, chalk, camera tape, and have been known to include a supply of the director of photography's favourite cigarettes. In his 'diddy-bag' are screwdrivers, pliers, alan keys, and other tools, balls of string, cans of oil, cleaning liquid and dulling spray, a depth of field calculator and a pad of time sheets—possibly also an exposure meter, just in case.

A young clapper loader lives in the hope that the focus assistant will fall ill, or a second unit be made up so that, having demonstrated his sense of responsibility in the way he handles the film stock, the very life blood of film making, he may proceed one rung further up the ladder.

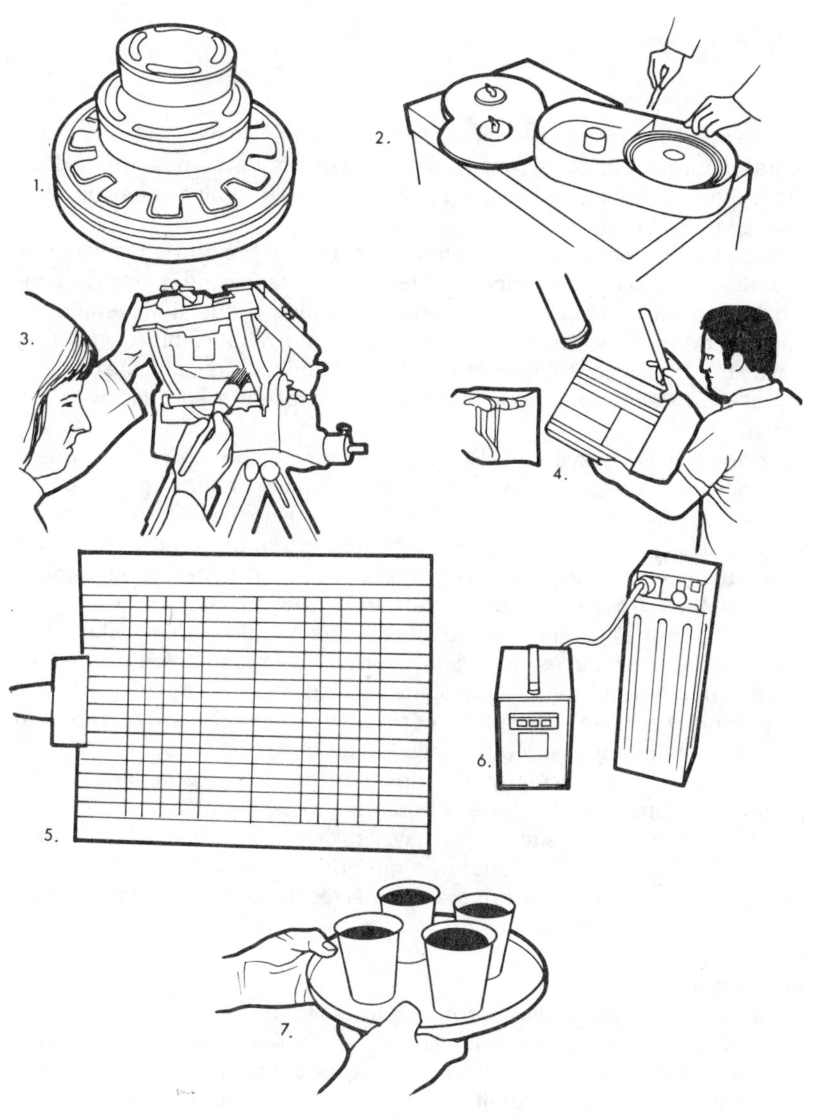

DUTIES OF THE CLAPPER LOADER

1. Ensure that an adequate supply of filmstock is to hand; 2. Load the magazines; 3. Operate the clapper board and announce the details on the beginning or the end of each take; 4. Keep a record of every take and prepare film for despatch to lab; 5. Make sure camera batteries are fully charged; 6. Service equipment after use; 7. Attend to all the needs of the cameraman.

*Everywhere, at all times.*

# Grips

To be a good grip is to be a highly skilled film technician who will have spent years learning his craft. Anyone who thinks a grip is just someone who does all the lifting, carrying and pushing does not know the value of having a top class grip on the unit.

When 'in shot' camera movements are required it is a grip who pushes or operates the dolly, jib or crane. In these circumstances he is actually contributing to the creative act of film-making. Major productions employ at least two grips, a key grip who is an 'organiser' and responsible for ensuring that all the artisan contributors are on hand when required and the camera, or dolly grip who confines himself to work closely associated with the camera.

A good grip is always busy. He physically moves the camera equipment from one position to the next; sets up the tripod or other form of camera support as called for and helps the crew set up the head and the camera. If a tracking shot is called for he ensures that there is a smooth and level surface for the dolly or crane to move over, makes up special camera mountings to go on anything from a horse drawn carriage to a racing car, puts down chalk marks or pieces of camera tape to mark where people or things should be positioned and produces, apparently from nowhere, anything needed for one of a hundred reasons. A grip starts early in the morning to ensure that everything is to hand for the camera crew when they arrive and who finishes late having ensured that everything has been tidied up.

On small units particularly, the driver of the truck containing all the camera equipment (and possibly the camera crew) doubles as a grip when the unit is at a shooting site. A grip may also drive the 'insert car' (when fast tracking shots have to be done) or a motorised crane, while another grip operates the jib arm. Key grips are expected to arrange certain special effects. In short, they must be masters of many skills.

## Equipment

Grips have available deals, boards, and tracks for dolleys and cranes to run on, '2–4–6's, wedges, boxes and pancakes to raise the level of things, hammers and screwdrivers, nails and screws, bits of wood and saws, ingenuity and good humour. If, in addition, he is a natural hustler, he can save the unit hours or days on a schedule.

On large productions he may work alongside carpenters (chippies), prop men, construction men, drivers, painters, and plasterers.

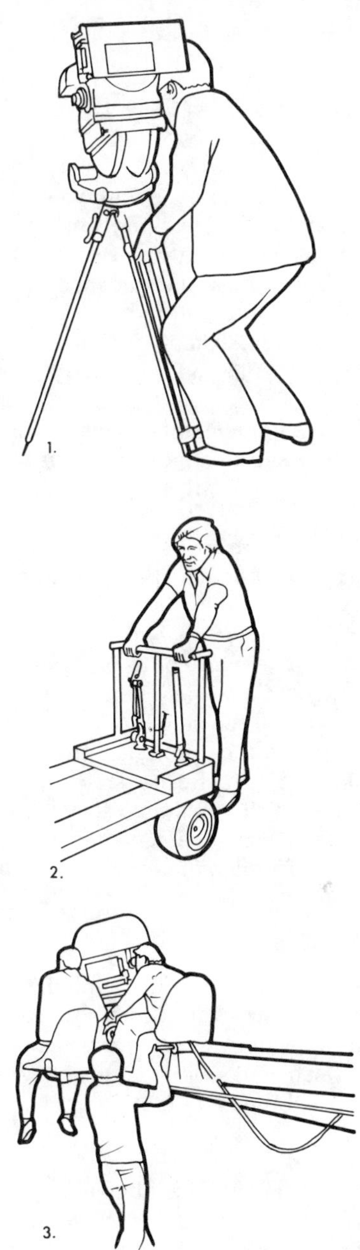


DUTIES OF THE GRIP

1. It is part of the job of the grip to ensure that the camera equipment is positioned as required; 2. When he pushes a dolly or handles a crane arm he is contributing to the in-shot operation of the camera; 3. When he operates a large crane with a camera crew seated on it he has the added responsibility of the personal safety of the crew to contend with.

# TV Film Cameraman

No matter how much electronic camera equipment is used for programme origination by any TV station, there still remains a vast amount of visual realisation which can most conveniently and economically be recorded on motion picture film for subsequent TV transmission. Documentary programmes, news items, inserts into programmes which have been partially recorded by electronic cameras, distant, difficult or foreign locations, etc. are all typical occasions where film may well be preferred to magnetic tape as the means of recording. In many situations, TV film cameramen are called upon to use portable lightweight cableless video cameras in much the same way as they are used to handling film cameras.

## Scope

Above all, the TV cameraman must be versatile in his approach to film making. Within a short space of time he can find himself shooting subjects as diverse as local news items, film inserts for drama or comedy programmes, a war in some far off foreign land or perhaps wild life in a jungle, remote from civilisation—and then come back again to filming a local celebrity opening a charity fête. It is an exciting and varied life for those who have the skill, the aptitude and the stamina.

Because the large TV networks commission programmes to be made on almost every conceivable subject there are good opportunities for a TV film cameraman to become a specialist in filming activities or sports in which he is particularly interested. For instance, a cameraman interested in motor sport might well find himself filming grand prix motor races at international venues.

## Crewing

Crewing varies widely in TV work. Bearing in mind that most film for TV is shot with 16mm cameras, a two-man (cameraman and assistant) crew is most often considered to be ideal (except for newsreel coverage which is always a solo effort). He also works with a sound crew and electricians who install, fix or hold the (often modest) lighting used.

## Equipment

A TV film cameraman may have to use almost every conceivable type of camera equipment, from the most sophisticated 35mm hand-held silent reflexes to the simplest 16mm gear. He may even use a super 8 camera when making reportage films in countries inhospitable to foreign TV crews but happy to pass or even encourage a 'tourist' with an amateur camera.

TV FILM CAMERAMAN

A cameraman who films for TV may often travel abroad to film subjects in which he takes a personal interest: 1. Cities; 2. Skiing.

*Mini-features, shot in one day.*

# TV Commercial Film Cameraman

TV advertising commercials are one of the most highly developed forms of cinematic art. They must be sufficiently attractive to catch and hold the eye yet serve the definite purpose of passing a message all within a period of 15, 30 or 60 seconds.

Discipline is most demanding. Every shot has to be exact, perfect and done without fail. Although some commercials are shot 'off-the-cuff' most are tightly scripted with precise storyboards that may have been the subject of meetings and conferences beforehand. Sometimes the artists compositions are cinematically almost impossible, yet that (and only that) is what the client has approved. Timings are extremely critical too. To be half a second out on each shot of a commercial may make it impossible to cut.

## Cinematic opportunities

For the creative cinematographer, commercials offer opportunities which may not occur in other fields. Commercial producers are hungry for new eye-catching ideas. Young cinematographers use the medium to learn and experiment. Older and more experienced directors of photography gain satisfaction from having 'time' to achieve perfection without the need to proceed quickly from one scene to the next as in feature work.

Even within the compact world of commercial production, cameramen are cast according to their speciality, bearing in mind the subject and visual effect required. A cameraman who has an eye for wildly off-beat effects may not be the man to create the delicate tonal range needed to suggest the softness of hair shampoo.

It is a most competitive business. The director of photography must satisfy the director, who must satisfy the producer of the TV production company, who, in turn, must satisfy an advertising agency anxious to please and retain its clients.

On top budget commercials only the best is good enough. So wherever possible, 35mm register-pin cameras are used. They give not only slightly improved definition, but any optical work and superimposition can be done better without one image 'floating' in relation to the other. The expression 'that will do' is not that of the commercial film maker.

TV COMMERCIAL FILM CAMERAMAN

Filming a 'cat food' commercial for TV.

*The cameraman who trades in tricks.*

# Special Effects Cameraman

Truly a 'specialist', the cameraman who specialises in trick photography and the photography of tricks will find employment on many productions as the leader of a select group of technicians who put 'production value' into a motion picture.

## Ways and means

A special effects cameraman may be called upon to produce solid or semi-transparent ghosts, displace or replace parts of an image, add to, or subtract from, a scene, combine a real scene with a model, create fires, explosions, earthquakes or ships sinking in heavy seas, photograph miniatures to look believably like the real thing, and translate into film images the combined imaginations of the producer, the author and the director—all at a small fraction of the cost of reality.

There are often many means to achieve similar results. The special effects cameraman must not only know all the different techniques available to him but also their cost-effectiveness, for the object is not only to create that which cannot be achieved by straightforward cinematography but also to save costs. It often happens that a screenplay calls for scenes which would be so expensive to create in reality that without the skills of the special effects cameraman the picture could not possibly be made.

## Techniques and the camera equipment

As many of the processes used for shooting special effects involve multiple exposures in either the camera or the printer, a register pin camera producing images of exceptional steadiness must be used. It must accept a multiplicity of filters and mattes in front of, or behind the lens, and allow adjustment of the shutter while the camera is running, and also permit filming at different speeds or in reverse. Camera mountings that may be rotated about the optical axis, or panned and tilted about the front nodal point of the lens, may also be required.

In a major studio, the special effects cameraman can use back-and-front projection and travelling matte work. Above all, the special effects cameraman must know the scope of the optical printer and of bipack printing. Some major studios have their own special effects printers but most often he must work in close co-operation with an appropriately equipped laboratory from the inception of the shot to the acceptance of the length of celluloid.

Some special effects shots may be completed in minutes, others take days, weeks or even months ... of patience.

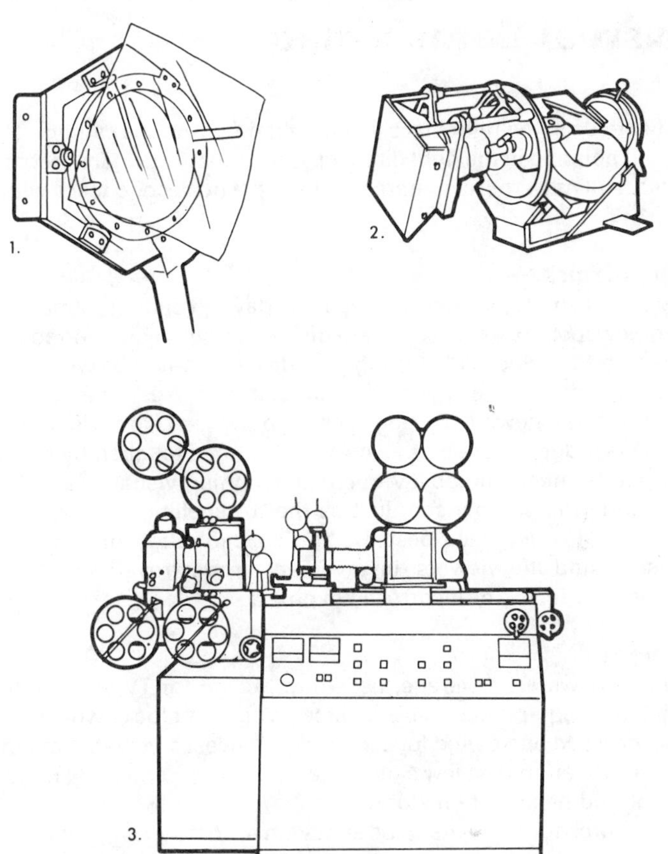


SPECIAL EFFECTS HARDWARE

The equipment used by the special effects cameraman can range from a simple distorting filter (1), through camera turn-over mounts (2), to sophisticated special effects optical printers (3).

# Newsreel Cameraman

The ability to be in the right place at the right time with the camera to hand, loaded and pointing in the right direction is the most important prerequisite of a successful newsreel cameraman. There are no second takes with hard news.

## Current affairs

The newsreel cameraman should begin his day by reading as many of the morning newspapers as he can manage, and should listen to radio news bulletins as often as possible through the day. He should be well informed on current affairs for, in reality, he is a journalist who uses a cine camera and a microphone as a newspaperman uses a typewriter. Like his literary counterpart, he is judged more by the story his pictures tell than by his skill in recording it. He needs no ability to create stunning visual effects.

To continue the comparison, just as the misspelling of a word or the misphrasing of a sentence destroys the concentration of a newspaper reader, so would the viewers' involvement be destroyed should a vital scene be out of focus, or incorrectly exposed.

## Equipment

Most filmed newsreel coverage these days, being for TV presentation, is shot on 16mm equipment using reversal type film stocks which may be quickly processed and edited for screening, if necessary, within minutes of arrival at a TV station. The few cinema newsreels still remaining use 35mm equipment and neg/pos film stock.

When pictures for news are to be shot with accompanying sound, single system (com-mag) cameras are most often used. Such cameras ensure that no time is lost synchronising picture to sound at the editing stage prior to presentation. Alternatively, a newsreel cameraman may be called upon to use portable video ENG (Electronic News Gathering) equipment for on site recording or beaming directly back to the TV station. Newsreel equipment must be lightweight, rugged and simple. As a one-man operator the newsreel cameraman is very concerned about the weight of each item and its carrying case, for he personally must carry it everywhere.

Because TV news must be almost instantaneous in its presentation, TV stations often rely upon associated stations, news film agencies, and local freelance cameramen (stringers) for much of their out of town news coverage, gathering it by the fastest possible means of transport and by all the national and international networks and satellite facilities available.

**Hours ahead of (+) or behind (−) Greenwich Mean Time. Certain countries use daylight saving time (1 hour during their summer) but no account of this is taken in the information below.**

|  |  |
|---|---|
| **AFGHANISTAN** | +4½ |
| **ALGERIA** | **GMT** |
| **ARGENTINA** | **−3** |
| **AUSTRALIA:** | |
| *NSW, Queensland,* | *+10* |
| *Tasmania, Victoria* | *+10* |
| *South Australia* | |
| *Northern Territory* | *+9½* |
| *West Australia* | *+8* |
| **AUSTRIA** | +1 |
| **AZORES** | −1 |
| **BAHAMAS** | −5 |
| **BANGLADESH** | +6 |
| **BELGIUM** | +1 |
| **BOLIVIA** | −4 |
| **BRAZIL:** | |
| *East* | *−3* |
| *West* | *−4* |
| *Acre* | *−5* |
| **BULGARIA** | +2 |
| **BURMA** | +6½ |
| **CAMBODIA** | +7 |
| **CANADA:** | |
| *Newfoundland* | *−3½* |
| *Atlantic* | *−4* |
| *Eastern* | *−5* |
| *Central* | *−6* |
| *Mountain* | *−7* |
| *Pacific* | *−8* |
| *Yukon ex Dawson* | *−8* |
| *Dawson City* | *−9* |
| **CANARY IS** | **GMT** |
| **CEYLON** | +5½ |
| **CHILE** | −4 |
| **CHINA (Peking)** | +8 |
| **COLOMBIA** | −5 |
| **COSTA RICA** | −6 |
| **CUBA** | −5 |
| **CYPRUS** | +2 |
| **CZECHOSLOVAKIA** | +1 |
| **DENMARK** | +1 |
| **ECUADOR** | −5 |
| **EGYPT** | +2 |
| **ETHIOPIA** | +3 |
| **FIJI** | +12 |
| **FINLAND** | +2 |
| **FRANCE** | +1 |
| **GAMBIA** | **GMT** |
| **GERMANY (E & W)** | +1 |
| **GHANA** | **GMT** |
| **GREECE** | +2 |
| **GUAM** | +11 |
| **HONG KONG** | +8 |
| **HUNGARY** | +1 |
| **ICELAND** | **GMT** |
| **INDIA** | +5½ |
| **INDONESIA** | +7 |
| **IRAN** | +3½ |
| **IRAQ** | +3 |
| **IRELAND** | **GMT** |
| **ISRAEL** | +2 |
| **ITALY** | +1 |
| **JAMAICA** | −5 |
| **JAPAN** | +9 |
| **JORDAN** | +2 |
| **KENYA** | +3 |
| **KOREA** | +9 |
| **LAOS** | +7 |
| **LEBANON** | +2 |
| **MALAWI** | +2 |
| **MALAYSIA** | |
| *West* | *+7½* |
| *Sarawak, Sabah* | *+8* |
| **MEXICO** | −6 |
| **MOROCCO** | **GMT** |
| **MOZAMBIQUE** | +2 |
| **NEW CALEDONIA** | +11 |
| **NAURU ISLAND** | +11½ |
| **NEW GUINEA** | +10 |
| **NEPAL** | +5⅔ |
| **NETHERLANDS** | +1 |
| **NEW HEBRIDES** | +12 |
| **NEW ZEALAND** | +12 |
| **NIGERIA** | +1 |
| **NORWAY** | +1 |
| **OKINAWA** | +9 |
| **PAKISTAN** | +5 |
| **PANAMA** | −5 |
| **PAPUA** | +10 |
| **PARAGUAY** | −4 |
| **PERU** | −5 |
| **PHILIPPINES** | +8 |
| **POLAND** | +1 |
| **PORTUGAL** | +1 |
| **PORTUGUESE GUINEA** | −1 |
| **RHODESIA** | +2 |
| **SAUDI ARABIA** | +3 |
| **SIERRA LEONE** | **GMT** |
| **SINGAPORE** | +7½ |
| **SOUTH AFRICA** | +2 |
| **SPAIN** | +1 |
| **SRI LANKA** | +5½ |
| **SUDAN** | +2 |
| **SWEDEN** | +1 |
| **SWITZERLAND** | +1 |
| **SYRIA** | +2 |
| **TAHITI** | −10 |
| **TAIWAN** | +8 |
| **TANZANIA** | +3 |
| **THAILAND** | +7 |
| **TRINIDAD AND TOBAGO** | −4 |
| **TUNISIA** | +1 |
| **TURKEY** | +2 |
| **UGANDA** | +3 |
| **USSR:** | |
| *Moscow* | *+3* |
| *Omsk* | *+6* |
| *Vladivostock* | *+10* |
| **UK** | **GMT** |
| **USA:** | |
| *Eastern* | *−5* |
| *Central* | *−6* |
| *Mountain* | *−7* |
| *Pacific* | *−8* |
| *Alaska-East* | *−8* |
| *Alaska-West* | *−11* |
| *Hawaii* | *−10* |
| **URUGUAY** | −3 |
| **VENEZUELA** | −4 |
| **VIETNAM** | +8 |
| **YUGOSLAVIA** | +1 |
| **ZAMBIA** | +2 |

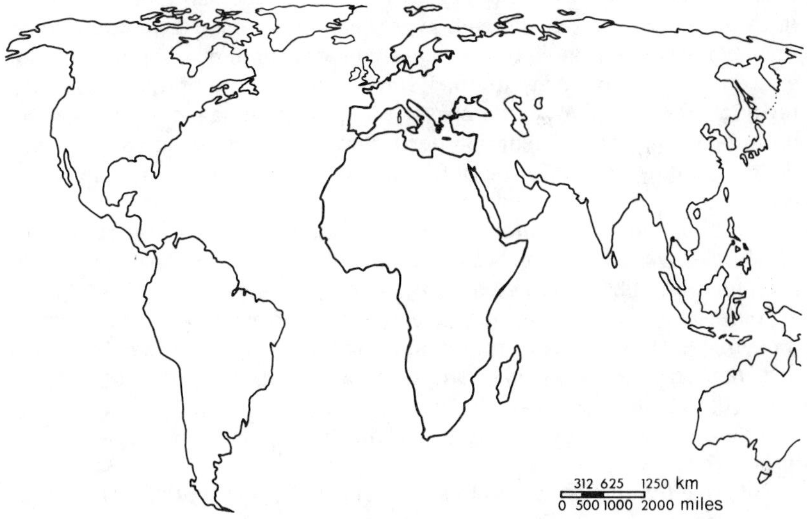


WHERE HE WORKS

The beat of the newsreel cameraman.

# Documentary Cameraman

The term 'documentary cameraman' covers a wide range of activities, from the specialist cinematographer who shoots films about people, places and things to the specialist scientist who uses film as a research tool, or to illustrate a thesis.

## Scope and equipment

Industry is one of the largest patrons of the documentary film maker. Specialised companies and 'in plant' film makers are employed to produce films about the work or processes of companies for informational, research and promotional purposes.

Naturalist film makers travel to the furthest corners of the world to expose and record the flora, fauna and topography of our environment and even beyond. Sports specialists, usually sports fanatics, make films about sporting happenings for fellow sportsmen. Medical practitioners record physical forms and operations and engineers use cinematography for research.

Increasingly, Super 8 is being used for documentary film making on the grounds of economy, as more sophisticated equipment, materials and laboratory and dubbing facilities become available for this gauge.

The specialist documentary film maker may call upon a range of equipment and materials far wider than that required by the cinematographer who works only for the cinema and TV. High speed cameras with frame rates in the hundreds or thousands per second; time lapse cameras which may expose only a single frame or less, per day; high power microscopes; endoscopes which reach into the depths of the human body; X-ray, infrared and ultra violet light materials and equipment; image intensifiers which permit photography in near darkness; Strobe light and cold lights, underwater and outer space forms of transport, etc—all these and many more, are the tools of the documentary cinematographer.

Very often the documentary cameraman prefers or is required to edit and finish his own film all the way through to the final presentation. At the premiere, he may also have to operate the projector.

There are no limits to the ingenuity that this type of work calls for, much of it beyond the realms of pure cinematography, as in devising say, a method to synchronise an 8 million frames-per-second camera to photograph a 500,000 volt electric spark.

Other aspects of documentary film making call for more of an artistic approach.

Only one thing is certain in this field of activity and that is that the man who does it, rarely finds his working hours repetitive.

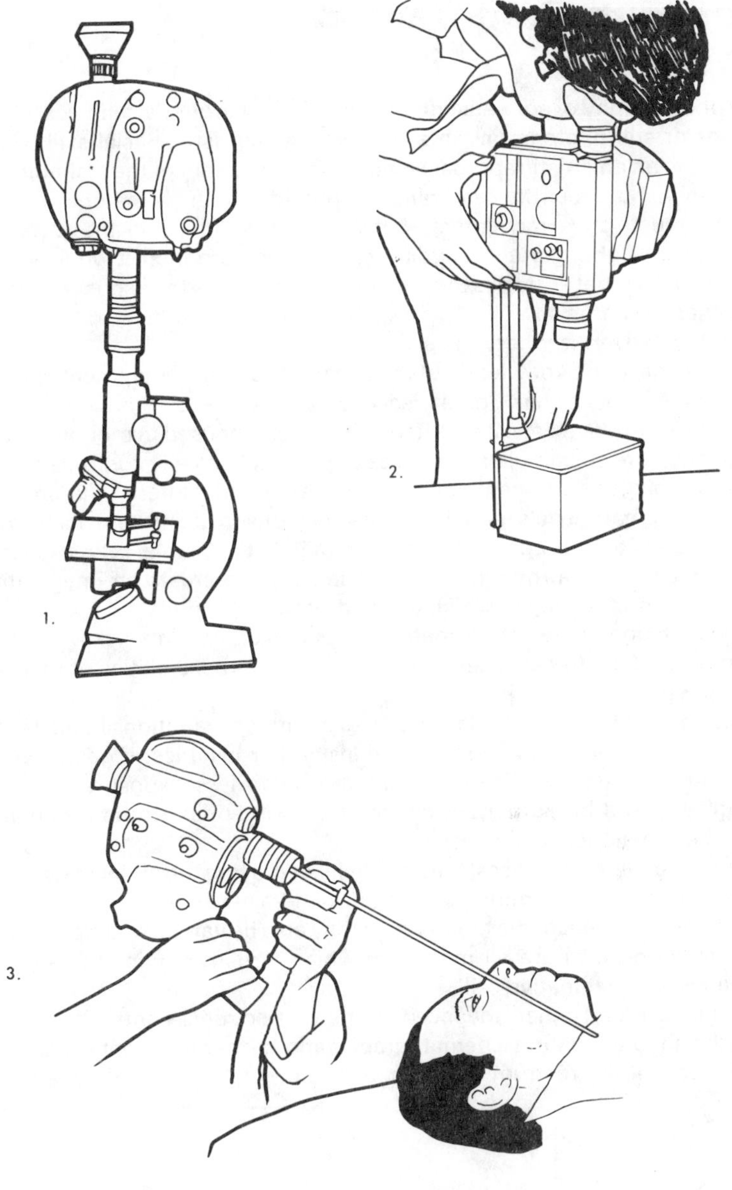


DOCUMENTARY AND SCIENTIFIC WORK

Typical equipment used in medical documentary cinematography.
1. Micro cinematography; 2. Macro cinematography; 3. Endoscopic cinematography.

*The cameraman who makes the inanimate move.*

# Rostrum Cameraman

The photography of animated drawings, title backgrounds, technical illustrations or still photography in a creative manner is a special skill which relies less on an artistic approach and more on mastery of the technology.

Given that the rostrum cameraman is provided with his material already drawn, the relative movements between the camera and the work are planned, his lighting pre-set, and his exposure determined, it would seem, at first glance, that all he has to do is alternatively press the exposure button and change the gels.

That, however, is where the skill begins.

With a thorough knowledge of the capabilities of his equipment and the filmstock he uses he has a great deal of control, and a contribution to make to the final product. Few, if any, items of equipment used in motion picture production are so complex or have as many mechanically controlled creative possibilities as the animation stand, table and camera. The photography of animation gels in their relative positions to illustrate say a historical section of a TV programme, where film is not available, is an aspect of work which gives a rostrum cameraman scope to create an imaginative presentation of a subject which might otherwise be dull.

A knowledge of the sensitometric characters of the filmstocks used and the means of controlling them add to the creative possibilities of rostrum camerawork.

Where double and multiple exposures are made, variations in the levels of exposure produce differences in density. For instance, if gels bearing drawings of shadows, fire or smoke are given less exposure than the foreground and background drawings they will appear to be less dense than the remainder of the scene.

The use of different lenses, the possibility of moving the optical axis of the lens relative to the aperture, tracking the camera in and out, all in combination with movement of the table (which may also be varied for different gels of a single picture) are all means by which a rostrum cameraman may make his an imaginative skill.

Computers and other advanced forms of electronic control are being introduced to animation cinematography and further highlight the technical aspects of the rostrum cameraman's role.

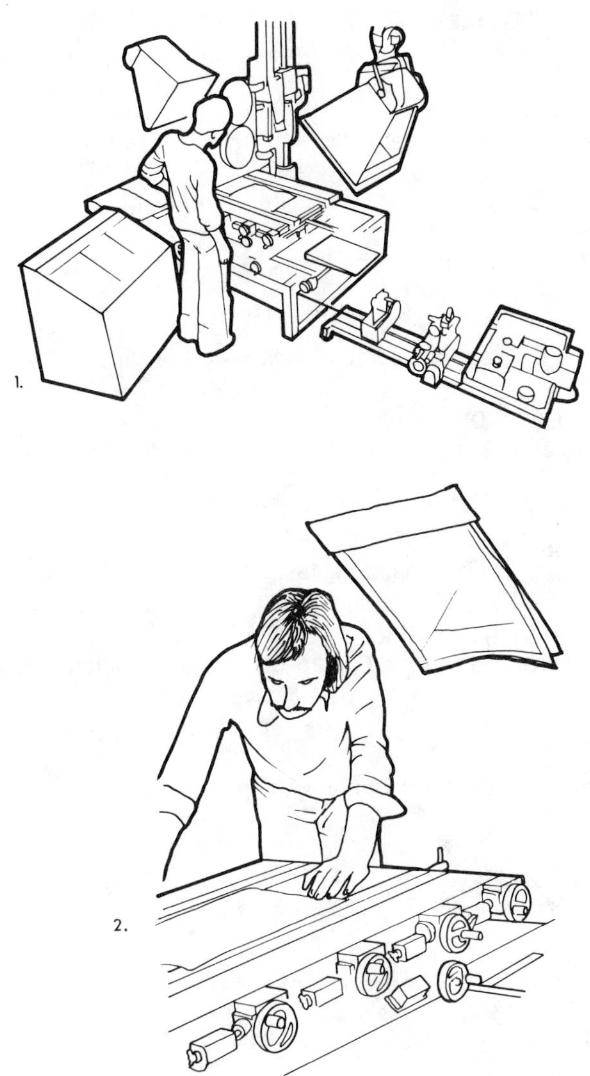

ROSTRUM CAMERAWORK

The work of the rostrum cameraman involves filming animation films, titles, photographs and drawings etc, (1, 2), much of it very complicated.

# Further reading


BADDELEY, HUGH W.
The Technique of Documentary Film Production (4th edition). 1975, Focal Press.

DALEY, KEN
Basic Film Technique. 1980, Focal Press.

HAPPE, BERNARD
Your Film and the Lab (2nd edition). 1983, Focal Press.

MILLERSON, GERALD
The Technique of Lighting for Television and Motion Pictures (2nd edition). 1982, Focal Press.

MILLERSON, GERALD
TV Lighting Methods (2nd edition). 1982, Focal Press.

RAIMONDO SOUTO, H. MARIO
The Technique of the Motion Picture Camera (4th edition). 1982, Focal Press.

SAMUELSON, DAVID W.
Motion Picture Camera and Lighting Equipment (2nd edition). 1986, Focal Press.

SAMUELSON, DAVID W.
Motion Picture Camera Data. 1978, Focal Press.